AF342380

Oklahoma Horizons Series

A HISTORY OF THE STATE FAIR OF OKLAHOMA

by
Bob L. Blackburn
and
Paul B. Strasbaugh

Series Editor: Kenny A. Franks

Published for the Oklahoma Heritage Association
by Western Heritage Books

ISBN: 0-86546-089-2

FOREWORD

This book, in our opinion, is an important step in the continued development of the State Fair of Oklahoma. After all, if we do not understand the past, how can we plan for the future. We think this book will provide essential information that can be used by the board, the staff, and everyone involved in the fair to set and then reach greater goals.

We also are pleased to share this important story with the general public, the very people served by the State Fair of Oklahoma. We want the people of this state to understand the fair and the dynamics that make it work. Together, with a greater understanding of both operational challenges and public expectations, all of us can work towards a common goal that helps everyone.

In the larger context, this book points out important themes that we should all remember. After reading about the commitment of Henry Overholser, we are struck by his vision, his determination, and his ability to dream big dreams and then pull together the resources to make it a reality. Like Overholser, we have to always stress the motto, "invest in the future."

Another theme that surfaces time and again in the book is the importance of adaptability. In a world where the only constant is change, we have to learn to recognize both problems and opportunities, and then do something about them. We cannot be locked into the past; we have to look ahead and chart our future.

If we on the board of directors can live by these guidelines, and if the general public understands the grand mission and potential of the fair, we are certain that the future is bright. We think this book will help achieve both of those objectives.

Donald J. Hotz   Edward Joullian, III

PREFACE

When this project began, we posed the basic question that any writer asks himself. Who are we writing for, and what could we emphasize to attract the widest possible readership?

The first part was easy. We wanted to write a history that would trace the development of the fair from the earliest days to the present time and identify the major themes and trends for the reader. The second part, discovering the general public's impressions of the fair, was a bit more complex.

We found that some people think of livestock competition and farm machinery, or perhaps the car show with the latest models. Others remember the flood of information found at the endless rows of exhibitors, especially the Made in Oklahoma exhibits with free samples of food. Most recall a montage of special moments—the color and sound of the midway, the monorail passing overhead, or the cowboy who fought to stay on a bull for the full eight seconds. Maybe it was a particularly good corn dog.

The State Fair of Oklahoma, we determined, is all of this and more, an institution that has served the state and its people for more than ninety years. But how did it happen to include such a diversity of attractions? And how did the fair grow and evolve into such an important part of the economic and cultural fabric of society?

In the course of research, we discovered a number of dominant themes that serve as connecting threads from the 1890s to the present time. One of the most dramatic is the boom and bust cycle of rapid growth followed by periods of troubled consolidation and momentary retreat. The State Fair, following periods of optimistic expansion, moved perilously close to insolvency in 1908, 1917, 1937, and again in 1954. Each time, the leadership of the city and state rallied to the cause and kept the institution alive and ready for the next upward cycle of investment and popularity.

The role of leadership is another theme that dominates the story. Beginning with C.G. "Gristmill" Jones and Henry Overholser, there has been a tradition of men and women who step forward at critical times to articulate problems and formulate solutions that carry the fair to the next level of success. That tradition can be traced through each decade to the present time and the accomplishments of community leaders such as Edward L. Gaylord, John Parsons, Bill Swisher, and Edward C. Joullian, III.

It is impossible to talk about leadership and the fair without weaving into the story the role of the Oklahoma City Chamber of Commerce. The Chamber created the fair, then nurtured its offspring through good times and bad. In 1917, when a bond issue was necessary to keep the fair alive, the Chamber took the lead. When a cash flow crisis threatened the fair in the 1920s, Chamber members pledged their resources to guarantee loans. In the 1940s the Chamber and its dynamic leader, Stanley Draper, Sr., paved the way to an expanded site and a brighter future. It is safe to say that there probably would be no State Fair of Oklahoma without the Oklahoma City Chamber of Commerce.

Another important partner has been the City of Oklahoma City. Since 1917, when voters approved a bond issue to rescue the fair from financial ruin, the City has owned the fairgrounds property. Today, the fair association, a nonprofit corporation, manages the actual planning and operation of the fair as well as management of the interim operations the other eleven months of the year. Although the balance of responsibility has changed periodically, and although there have been occasional disagreements over policy, the relationship has remained strong and mutually supportive.

As the research progressed on the book, and we determined what happened and who was involved, the next questions were obvious. Why did the State Fair of Oklahoma attract and maintain this level of commitment? And what motivation drove business leaders, educators, and the general public to invest so much time and money into one institution that initially lasted only ten days a year? The answers were provided by the evidence. It was a combination of economic development, education, and entertainment that made the first fair a success; the same formula keeps the fair moving forward today.

The State Fair of Oklahoma has always been one of the urban booster's most efficient tools for economic development. In 1907, when the state was predominantly rural and Oklahoma City was a struggling community of 30,000, the fair attracted crowds of consumers to town and provided an efficient opportunity for manufacturers, wholesalers, and retailers to spread the word about their products. In an era without electronic media and limited circulation of newspapers, this annual rite of reaching a large market was exceptionally important.

The marketing value of the fair remains a major weapon in the war for economic development. Major corporations compete for the most strategic spaces, booths sell out early, and a wide variety of entities, from nonprofit service groups to state and federal agencies, line up to come face to face with an army of potential customers. No other event or place provides such efficient exposure.

Then there is the economic benefit of getting so many people into Oklahoma City for extended visits. The annual fall fair alone attracts 1.6 million people, while the fairgrounds is booked with 1,600 event days a year, including more than a dozen national or international horse shows that have made Oklahoma City the horse show capital of the world. Together, the events held at the fairgrounds create an annual economic impact of $200 million.

An added benefit of the fair is that all of this activity is wrapped in the larger cloak of education. In 1907, as in 1994, information was power, and the fair provided the perfect vehicle to get the latest and the best information to the general public. Whether it was a display on mineral resources of the state or an exhibit on radios, the fair was and still is an efficient method of communication.

The educational value of the fair has gone far beyond the simple presentation of information. The fair has been used repeatedly as a way to improve society, whether it was a seminar on increasing dairy production or a competition honoring the outstanding educator in the state. The fair, with its vast marketing potential and availability of space, has made a real impact on educating both the youth and general public of Oklahoma.

Neither economic development nor education would be possible without the third element of the fair's success: entertainment. In 1907, after the premiums, Gristmill Jones allocated the largest budget amounts to military bands, horse racing, recitations, and special events that would attract the crowds. Later managers turned to aviation, automobile racing, ice dancing, and rodeos—whatever the public demanded—to keep the turnstiles turning.

One element of entertainment that has not changed is the midway. Originally called Sooner's Trail, the midway adds color and excitement to the fair, provides a means for having fun, and generates income that supports the educational and competitive components of the total program. The midway, like other forms of entertainment, is the financial fuel that drives the machine that is the State Fair of Oklahoma.

This book would have been impossible without the support and assistance of many people. First in line are Don Hotz and Edward Joullian, III, who clearly understand the importance of history. As they said in initial meetings, a working knowledge of the fair's development, including mistakes as well as accomplishments, is essential if they are to chart a successful path to the future. Their keen understanding of the fair and the dynamics that affect its well being were critical in the direction of the research and writing.

Other members of the board and staff were especially helpful. Board members involved in the fair's development, such as Edward L. Gaylord, Bill Swisher, and John Parsons provided both their time and insights during the research. Staff members such as Francis Young, Scott Muntz, Pam Cook, and Judy Matheny were ever ready to help find that one additional piece of information necessary to tell the complete story. Most memorable was an extended interview with Orval O. "Sandy" Saunders, general manager of the fair from 1961 to 1985, who willingly and candidly shared his knowledge and understanding of the fair.

As any historian will testify, the success of a book depends on the availability of primary resource materials. In this project, we were blessed. Staff at the State Fair of Oklahoma have been saving newspaper clippings and archival documents since 1907. When combined with a full set of all board minutes since the organizational meeting in 1907, this archival treasure provided a wealth of information. Additional research was conducted in the newspaper and archival collections of the Oklahoma Historical Society, the Metropolitan Library of Oklahoma County, and the Western History Collections. Photographs were drawn largely from the State Fair Archives, the collections of the Oklahoma Historical Society, and the incomparable files of the *Daily Oklahoman*. Carol Campbell, head librarian at OPUBCO, was especially helpful in the search for photographs.

Finally, we want to dedicate this book to all the board members and staff who have toiled since 1907 to make the State Fair of Oklahoma one of the top events in the nation. Their efforts have done much for the material and cultural growth of the city and state. Our hats are off to those dedicated individuals.

Bob L. Blackburn   Paul B. Strasbaugh

CONTENTS

Foreword...vii
Preface...ix
1 A Territorial Tradition.................................3
2 A Fair is Born.......................................29
3 Investing in the Future.............................51
4 Another Chance.....................................79
5 The Big Move......................................111
6 Completing the Dream.............................137
7 All Things to All People..........................163
8 Expanding Horizons...............................189
  Directors, State Fair of Oklahoma, 1907-1994.............222
  Notes...225
  Index...231

A HISTORY OF THE STATE FAIR OF OKLAHOMA

Chapter One

A TERRITORIAL TRADITION

The downtown business district of Oklahoma City was unusually quiet for a Saturday morning, a time when farmers and shoppers normally filled the streets of the young boom town of 50,000 people. Even the suburban neighborhoods along the streetcar line as far north as 15th Street were relatively still on that fall morning of October 5, 1907. All attention, it turned out, was focused on the opening ceremonies of a new institution located a mile east of town—the State Fair of Oklahoma.

Fittingly, the featured speaker at the grand opening was Governor Charles N. Haskell, just elected to an office that would not exist for another month when Oklahoma officially would become the 46th state. Haskell, although a native of Oklahoma City's primary urban rival, Muskogee, praised the fair and the people who had made it a reality:

> Nothing can be devised that would serve better to illustrate the pride taken in Oklahoma City by the people of Oklahoma generally than this first state fair, held in Oklahoma City, and organized almost wholly by Oklahoma City people. No form of public movement so calls into play the sympathetic interest of all classes of society as does the successful state fair.

Opposite page: Economic performance, both on the farm and in the city, was the primary motivation behind the first fairs in Oklahoma City. Here, an exhibitor sits with a variety of produce at the 1898 Street Fair (Courtesy Oklahoma Historical Society).

Muskogee, founded in 1871, was the largest town in the territory and the site of the first fair in what would become Oklahoma (Courtesy Oklahoma Historical Society).

The needle work of the seamstress, the finished product of the mechanic's skill and the display of the merchant are here ranged side by side for the entertainment and enlightenment of the general public. In no other way can we be so thoroughly convinced of our dependence on one another. It proves to us that life is a mutual affair, and that in the framework of organized society there is a place and necessity for all classes.

Starting with that dedication ceremony, where Governor Haskell so eloquently recognized its many benefits, the State Fair of Oklahoma was launched on its course into history.

Looking at the success of the State Fair almost a century later, it might seem that its founders were either brilliant strategists or simply lucky enough to find a formula that worked. In fact, neither assumption is totally correct. First, the history of the fair would be filled with failures as well as successes. Luckily, the men and women behind the effort would have the persistence to keep going. And second, that first fair with its combination of agricultural displays and entertainment was no accident. Like any successful institution, the inaugural fair was built on a foundation of experience, a territorial tradition of education, celebration, and economic boosterism.

Culturally, Americans inherited a legacy of fairs dating to the late Middle Ages. Even in the darkest decades of the Medieval era, merchants and guilds hosted trade fairs to attract customers and feed the fires of economic development. Then there were the traditional fall feasts, in Latin called *faires*, that were community-wide celebrations of the harvest, a time for rural folk to socialize and share the bounty of the land. Gradually, such social events grew with dances, concerts, speeches, and other forms of entertainment that dismantled the walls of isolation, even if only for a few short days each fall.

That rich tradition was transplanted to the New World, and here it took root and blossomed. First came community fairs, with local competitors vying for bragging rights. As towns emerged from the wilderness, mercantile elements were added to encourage sales and recruit new businesses. By 1840 the tradition had grown to the point that both New York and New Jersey organized the first state fairs, a combination of the old harvest celebrations and the new economic boosterism of trade fairs.

It was Native Americans—more specifically the Cherokees—who held the first fair in what would become the state of Oklahoma. It was in 1845, only a few short years after the Cherokees' forced removal from Georgia, when the Agricultural Society of the Cherokee Nation staged a one-day fair near Tahlequah to promote stock raising and the planting of cash crops. Despite the remarkable achievements of the Indian nations during their golden age of tribal independence, the frontier society of early Oklahoma was still too young, too far removed from major markets, to support a broad-based fair.[1]

The turning point in both the history of Oklahoma's economic development and the growth of fairs was the coming of the railroad. In 1871 the Missouri-Kansas-

The first fairs held in the territory focused on Native Americans and their efforts to cope with a rapidly changing world (Courtesy University of Kansas Libraries).

The Oklahoma City Post Office as it looked on April 22, 1889, the day of the first land run opening the territory to non-Indians (Courtesy Oklahoma Historical Society).

Texas Railroad, better known as the KATY, laid the first steel rails through the Indian Territory following the old Texas Road. Even before construction crews had departed, towns were platted: Vinita and Wagoner in the Cherokee Nation; Atoka and Durant in the Choctaw Nation; and the largest town of all, Muskogee, founded in the Creek Nation at the point where the tracks crossed the Arkansas River.

The town of Muskogee grew quickly as mostly white merchants set up shop to serve the needs of full bloods, mixed bloods, and former slaves who suddenly had an outlet to distant markets for their cash crops and livestock. To encourage trade and economic development, a group of leaders in Muskogee organized the Indian International Fair in 1874. Their stated goal was "the advancement of the Indian people...by fostering industrial pursuits, the education of the people, and the progress of moral teachings." That first fair attracted about 5,000 visitors to a 160-acre site east of Muskogee with horse races, two exhibit buildings, and a short list of premiums.[2]

Within a year the General Council of the Indian Territory officially recognized the Muskogee Indian Fair as a "benefit to the agricultural, mechanical and stock growing interests of the Indian Nations" and encouraged tribal members to attend the second annual event scheduled for September 14-17, 1875. Stock to fund the fair was sold at $25 a share, and a board of directors was elected with representation allotted to the Cherokees, Creeks, Seminoles, Peorias, Ottawas, Choctaws, and Chickasaws. Pleasant Porter, chief of the Creeks, chaired the grand council.[3]

On July 4, 1889, the people of Oklahoma City gathered for a grand celebration in the Military Section east of the tracks. The day ended in tragedy when the makeshift grandstand collapsed under the weight of the crowd (Courtesy Oklahoma Historical Society).

This bird's-eye view of Oklahoma City in 1890, looking south towards the North Canadian River, was drawn less than one year after the land run. The 1892 territorial fair was located on the military reservation, east (left) of the tracks (Courtesy Oklahoma Historical Society).

The resulting intertribal fair began with a parade of bands and tribal groups, each carrying a banner that proudly set forth the ambitions of reformers and mixed-blood leaders: "Agriculture is the Source of Wealth" from the Cherokees; "The Result of Peace" from the Caddos and Apaches; and "We need Schools, Cows, and Ploughs" from the Kiowas. In addition to the standard fare of horse races, there were exhibits that featured Indian manufactures such as robes, furs, vests, blankets, moccasins, bows and arrows, buckskin dresses, and fine beadwork. By 1892 and the "Fifteenth Annual Indian International Fair," a long list of premiums attracted entries ranging from fruits and cooking to handiwork and territorial products. There was even a premium for the winner of a bicycle race.[4]

To that time, the Muskogee fair had little competition in the territory other than periodic horse races. A group of men in nearby Fort Gibson tried a fair, but they could not compete with their richer neighbor. Muskogee was the largest town in the future state, the seat of federal and tribal government, and an economic powerhouse built on the wealth of farmers, ranchers, and transportation in the Three Forks area. That dominance, however, did not mean there were no rivals, especially in terms of ambition and boosterism. Less than 90 miles to the west, in the recently opened Unassigned Lands, there was an upstart town with great expectations—Oklahoma City.

Boosterism was the birthright of Oklahoma City. Like

Captain Daniel F. Stiles, who helped organize the first fairs (From Harper's Weekly).

H. OVERHOLSER,
Oklahoma City.

Henry Overholser, listed in the city directory as a "capitalist," supported the territorial fairs through stock purchases and sponsorships (Courtesy Oklahoma Historical Society).

Guthrie, its urban rival to the north, Oklahoma City was founded in the greatest one-day boom in American history—the land run of 1889. On the morning of April 22, 1889, the Unassigned Lands was an island of unfenced grassland withheld from non-Indian ownership and untouched by the plow. At high noon that isolation ended when 50,000 land-hungry pioneers rushed headlong into history and claimed every part and parcel of the "promised land." By nightfall, 10,000 people were crowded onto the townsite of Oklahoma City, stretching from what would become North 7th street to South 7th street and from the Santa Fe tracks to Walker Avenue.

The people who settled the town brought with them their cultural baggage from the older states, but with a new twist fashioned from the experience of settling the great American West. Generally, they were gamblers in the sense that they were willing to risk time and resources for future profit. The fact that they had left the security of home and hearth to invest in a new frontier was proof of that. These rough-cut pioneers also expressed a deep and abiding faith in the powers of boosterism, the unfettered pursuit of economic development, and the benefits of education as the hope for tomorrow. Add to this cultural mixture a readiness to celebrate and socialize, and you had the perfect ingredients for that great American tradition, the fair.

In September of 1889, only five short months after the dust had settled from the land run, the people of Oklahoma City organized their first attempt at a fair, a one-day celebration that was, in reality, a salute to themselves. The day began with a parade of congressmen who "so nobly fought for the passage of the bill that opened this glorious land." After a lunch of barbecue, bread, and pickles, politicians ascended the speakers' stand at the

Round Grove and pledged their support for Statehood, each one "wildly applauded" in the words of one reporter.[5]

Then came the exhibits, divided into the Agricultural Hall and the Art Hall. As one town booster wrote, the agricultural display assembled by J.E. Sayre was a "small wilderness of agricultural wealth…, a collection of products of new soil well calculated to elicit applause from visitors and strengthen the faith of every farmer whose luck has allowed him a home in a land so fair." Considering the late planting season after the land run, the variety was impressive. There were stalks of corn that "tower well up toward the sixteen foot ceiling," cotton stalks that carried 207 bolls, watermelons that weighed as much as 72 pounds, sweet potatoes that averaged four pounds each, and a display of wild grasses gathered within a few miles of the city.[6]

The Art Hall next door was a "gratifying display of commercial and artistic merit," with groceries, millinery, hardware, and dry goods "showing at a glance the solidity, beauty and quality of the various articles of consumption and commerce in the metropolis of the great southwest." Local ladies added to the visual scene with paintings of dogs, cows, pastoral scenes, and Oklahoma landscapes. As one writer waxed, it was a "poem of color and beauty." It also was a chance for the people of Oklahoma City to pat themselves on the back, put forth an image of success, and stimulate the forces of free enterprise.[7]

Overholser's Grand Avenue Hotel (left) was the site of early merchants' fairs (Courtesy Oklahoma Historical Society).

Within four months of the land run, farmers were anxious to display their first harvest at a local celebration in Oklahoma City. This selection of vegetables was grown by farmer A.W. Bennett (Courtesy Oklahoma Historical Society).

The urge to combine celebration with economic development was strong during the territorial years, and any variety of events from the 4th of July to the arrival of politicians could generate a call for a carnival or fair. On April 22, 1890, the first anniversary of the land run, town leaders organized a "Merchants' Trades Carnival" to be staged in the large double storeroom in the Overholser Building.

More than 1,000 people gathered for food and a "grand march" of women, each wearing a costume featuring a different business. One of the walking advertisements was Miss Lela Hendrey, who represented the Anchor Drug Store with a white gown, bracelets of rubber bells tied with blue ribbons, a necklace of different colored vials, and a cap of yellow sponges ornamented with paint brushes, pen holders, and other sundries. Another model was her sister, Miss Belle Hendrey, who promoted the W.J. Pettee Hardware Company with a black dress featuring a dog chain belt from which dangled an "artistic confusion of door knobs, oil cans, fishing tackle, cork screws, pepper boxes, and other small articles."[8]

A more serious attempt to promote economic development was the annual poultry fair, first staged in 1890. In January of 1891 the second two-day exhibit attracted breeders from Kansas who hoped to sell stock and eggs, as well as farmers from the territory who sought any opportunity to make extra cash. As one writer recognized, the Poultry Fair and its promise of premiums "will

C.G. "Gristmill" Jones came to Oklahoma City in 1889 to become one of the town's founding fathers. He led the effort to organize and fund the territorial fairs of 1892 and 1894, and would later resurrect the idea in 1907 when economic conditions were more favorable (Courtesy Oklahoma Historical Society).

operate as a stimulant to poultry raisers to increased efforts in brooding birds of the choicest qualities."[9]

The desire to combine exhibits, competition, and entertainment in hopes of profit was not limited to poultry men or merchants. In 1891 Oklahoma City was teeming with hungry wholesalers looking to expand markets, aggressive bankers seeking depositors, and real estate promoters who equated population growth with the Holy Grail. It was not surprising then that the Commercial Club quickly came to the conclusion that the common good of the city—which meant the economic growth of the city—could be served with a fair, that tried and true tradition of the American frontier.

In the spring of 1891 the president of the Commercial Club appointed a new committee to pursue the possibilities of a combined fairgrounds andball park. They were to approach the local military commander, Captain Daniel F. Stiles, and seek permission to build facilities on the military reservation east of the townsite. Not surprisingly, leading the charge in this new initiative was the president of the Commercial Club, C.G. "Gristmill" Jones.[10]

A native of Cumberland County, Illinois, Jones was the proverbial city booster. He had come to Oklahoma City in January of 1890 to build the first flour mill, earning him the nickname of "Gristmill" Jones. With long-time colleague Henry Overholser, he invested in the "grand canal" during the bleak winter of '89 and '90, then turned his energies to railroad building and town development. He later courted and won the Frisco extension from Sapulpa to Oklahoma City, built the industrial belt line railroad around the city, and financed several trunk lines on his own, including the Oklahoma and Southwest Railway from Oklahoma City to Quanah, Texas. His public service career was just as impressive, including terms as mayor and the first state senator from Oklahoma City. The most enduring object of his interest throughout all of these accomplishments, however, was not profit, not politics, but rather the fair.[11]

When Jones appointed a committee to develop a fairgrounds and ballpark, he expressed his ultimate purpose in an open letter: "this city desires to further increase and more firmly establish the trade of those outside concerns which do a general business with our retail and wholesale merchants." Moreover, he wanted to "raise a fund for the Commercial Club to be used in...building rail and wagon roads and otherwise increase our commercial interests." To Jones and his fellow businessmen, what better way to accomplish both than to sponsor a territorial fair that would combine economic development with the lure of high society and entertainment.[12]

While negotiations were underway with the Department of the Interior for use of the military reservation, the committee recommended an immediate, more modest effort referred to as the "Commercial Club Fair and Exposition." It was to be held in the recently constructed Overholser Opera House, with a program that included raffles, sales booths, displays, and entertainment. Prominent on the planning committees were several men who would gain valuable experience—Ed Overholser, James Geary, Captain Daniel Stiles, and C.G. Jones— men who would eventually apply themselves to more ambitious efforts.[13]

When the fair opened on April 4, 1892, much of the credit was given to the ladies, who, as one reporter noted, "employed their time in converting the opera house into an Eden of beauty with its gaily colored stands and ornaments." The largest displays were sales tables where donated merchandise was sold as a fund raiser.

The "Premium List" of the Territorial Fair of 1892 included an impressive number of categories, such as photographs, penmanship, and a "doll's outfit by girl under 13." The premium book was sponsored by more than 100 local businesses, including the Grand Hotel, built and owned by Henry Overholser (State Fair Archives).

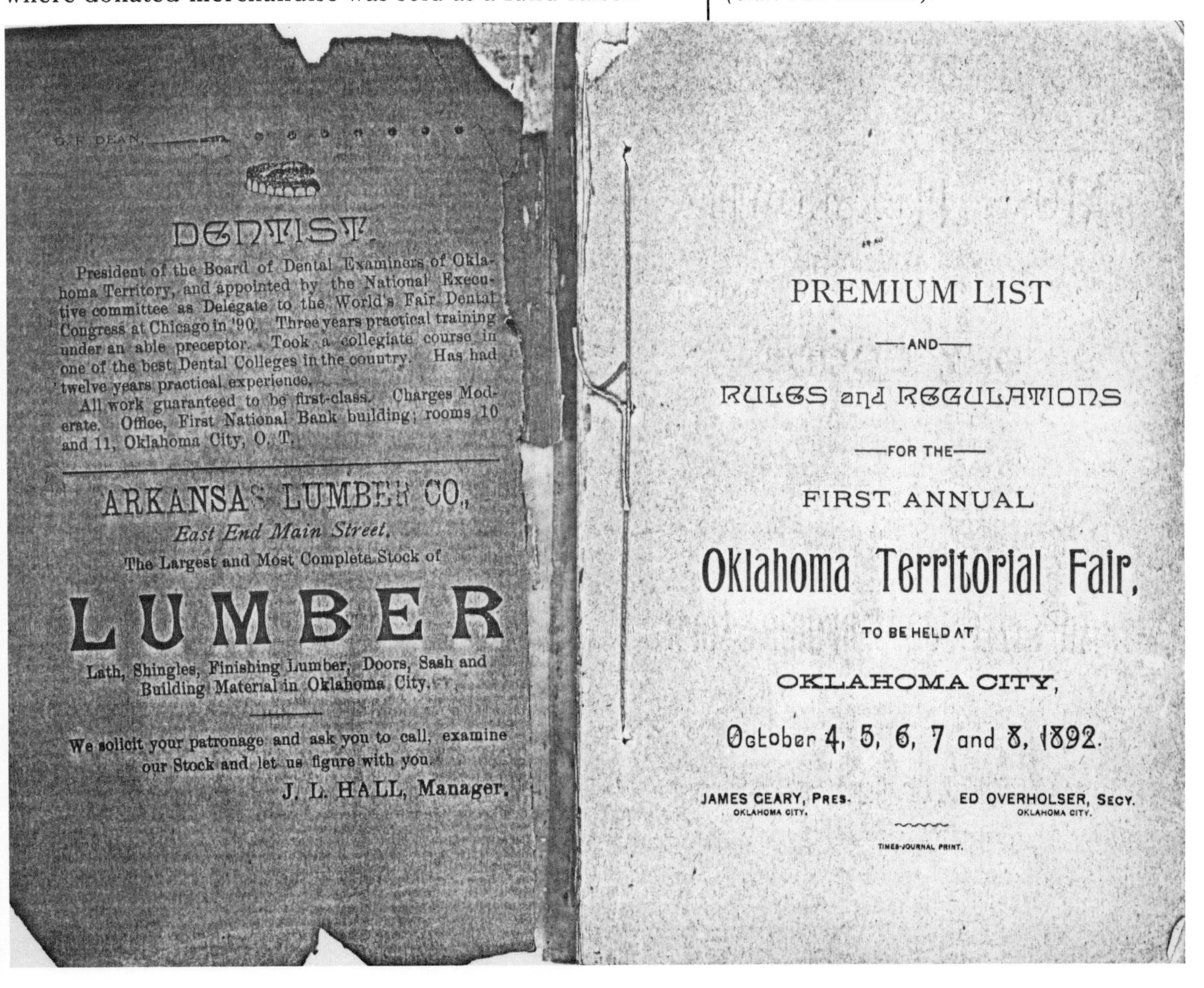

This float in the 1899 Street Fair, featuring a pioneer cabin and musical group, was sponsored by Ragon and Atwood Lumber Company. Behind is another float with samples of wallpaper on display (Courtesy Oklahoma Historical Society).

There was the Box Table, manned by eight ladies, and the Horseshoe Table, with an "effective body of drummers" where visitors could purchase everything from a "necktie to a chance on a $200 surrey." Rounding out the service booths were the Martha Washington Table, with lemonade and ices for sale, the Fancy Work Table, with crafts for sale, and the Lunch Table, offering a fare of coffee, cocoa, tea, and sandwiches.[14]

The east side of the display area was assigned to the brewers, who represented an important industry during the pre-prohibition territorial years. Included were the Ferd Helm Brewing Company, the Pabst Brewing Company, the Val Blats Brewing Company, and the Anheuser-Busch Company, with "an immense pyramid of their famous brands gaily ornamented with a profusion of flags and at the top a huge eagle." The exhibit hall also included displays of meats, lumber, farm implements, surreys and road carts, hardware, and stoves.[15]

Entertainment, as usual, was an important part of the exposition. Nightly speeches were made, led by Jones, Territorial Governor Abraham Jefferson Seay, and chairman of the fair Ben Miller. Recitations and musical offerings included instrumental and vocal solos as well as local bands and choral groups, who sang favorites like "The Lord is Great." Interspersed were raffles and good-natured voting on the "favorite girl" and the best floral arrangement.[16]

The results were mixed. Attendance averaged less

The arrival of the St. Louis and San Francisco Railway, better known as the Frisco, in 1898 was the symbolic beginning of an economic boom that propelled the population of Oklahoma City from 4,000 to 64,000 in little more than a decade (Courtesy Oklahoma County Metropolitan Library).

than 200 a day, and receipts from the ten-cent admission and sale of donated goods totaled little more than $1,000—but there were signs of promise. The exposition had attracted new customers from out of town, and the newspapers carried notes about prominent visitors in town for the big event. The Santa Fe and Choctaw railway companies reduced fares for the duration of the fair and ran excursions from towns such as El Reno and Guthrie. Clearly, the effort was not a financial success, but it was enough to convince city leaders that their interests could be served through fairs.[17]

On June 28, 1892, with the spring exposition fresh in their minds, members of the Commercial Club voted to organize a fair association. This time it would be larger, with permanent grounds and improvements and would include a greater role for farmers and stockmen. For capital expenses, they voted to issue 1,000 shares of stock priced at $10 a share. Several members immediately pledged to buy 200 shares, and within a month 66 different men in the community had purchased 600 shares. By August workers moved onto the south 80 acres of the military reservation and began construction.[18]

The constitution and bylaws, written by C.G. Jones, adopted the name "Oklahoma Territorial Fair Association" and stated the mission as "the development of agriculture, horticulture, stock-raising, mechanics, manufactures, household economy, and the encouragement of the fine arts." A board of directors, elected by the stockholders, would direct the association, with general management overseen by an executive committee consisting of officers and four members from the board of directors. The first executive committee consisted of James Geary, president, C.G. Jones, vice-president, F.M. Riley, trea-

The premium list from the 1892 Territorial Fair included a wide variety of categories (State Fair Archives).

PREMIUM LIST.   91

CLASS 7—Photographs.

No.		PREMIUM.
669.	Display of cabinet photographs, not less than fifty	1 00
670.	Display of large photographs, not to exceed 10	1 00
671.	Display of views, not more nor less than 20 ...	1 00
672.	Photograph portrait in water colors	1 00

CLASS 8—Penmanship.

673.	Best and largest collection of pen work, not to exceed eight different specimens	Diploma
674.	Specimen pen drawings	Diploma
675.	Specimen business writing	Diploma
676.	Specimens flourishing	Diploma
677.	Sample card making	Diploma

CLASS 9—Children's Department.

Exhibitors in this class are confined to boys and girls under 15 years of age.

678.	Best patchwork quilt	1 00
679.	Best sample of sewing	1 00
680.	Best sample of darning	1 00
681.	Kensington embroidery	1 00
682.	Outline on linen	1 00
683.	Carving in wood	1 00
684.	Collection of coins	1 00
685.	Doll's outfit, by girl under 13	1 00

No articles included in the following classes will be given a premium that have been exhibited more than once before. In classes 10, 11, 12, 13, 14, 15, 16, 17 and 18, all articles must be entered in the name of the maker. No premium allowed on articles purchased or borrowed.

The judges are instructed not to award premiums unless the work is completed and worthy of merit.

CLASS 10—Carpets and Rugs.

686.	Ten yards or more woolen carpet	1 00
687.	Ten yards or more rag carpet	1 00

surer, Ed Overholser, secretary, and members D.F. Stiles, William McClure, H.G. Trosper, and D.C. Richardson.[19]

It was an auspicious group. Geary, one of the most colorful, was a pioneer who had made the Pikes Peak gold rush in '59, fought in the Civil War, freighted for the frontier army, and worked in ranching enterprises with legendary showman William "Buffalo Bill" Cody. He had made the land run of '89 and opened the first bank in Oklahoma City. By the time he volunteered for the first fair, he was in the real estate business with plans to subdivide Oklahoma City's first fashionable residential neighborhood, Maywood Addition.[20]

Ed Overholser was the son of "city father" Henry Overholser, the major shareholder in the fair association. Ed was 23 years old and just beginning his career as manager of the Overholser Opera House. Years later he would be elected a county commissioner and even serve two terms as mayor from 1914 to 1918. Captain Stiles had gained prominence as a steadying force during the hectic days of the run and the extended period of martial law when provisional government collapsed.

William McClure was an '89er known for his ingenuity. On the day of the run, he had posted hired hands with fresh horses along the best route between the Kickapoo line and the future site of Oklahoma City. By changing horses every few miles, he was the first legal settler to reach Oklahoma City, where he claimed two lots downtown and a quarter section south across the river. Trosper and Richardson were '89ers who had done just as well. Trosper owned several farms and was doing a lively business in real estate while Richardson owned several businesses.[21]

Under the guidance of these city fathers, the fairgrounds quickly took shape as a crew of 50 men and 20 teams descended on the 80-acre site located just east of the Santa Fe tracks between 4th Street and the river.

Main Street, seen here looking west from Harvey (Courtesy Oklahoma Historical Society).

Ed Overholser, the son of Henry Overholser, earned a place in the history of the fair. As a young man, he served as secretary of the territorial fair, and in later years he was mayor when the city passed a bond issue paying off the fair's debts and transferring title of the fairgrounds to the city (Courtesy Oklahoma Historical Society).

*Anton Classen, an '89er who had settled
originally at Edmond, moved to Oklahoma City in 1897. He invested heavily
in real estate, then became active in the
reorganized Chamber of Commerce,
which supported the steet fairs of 1898
and 1899 (Courtesy Oklahoma Historical Society).*

They erected a fence, drained the swampy land in the bottoms, and cleared an area for a half-mile race track and grandstand. A crew of 18 workers started on the Exhibit Building, which was 500 feet long and 24 feet wide, large enough for 100 box stalls. Nearby was the Floral Building, designed in the form of a cross, with each two-story section measuring 50 feet by 24 feet. The final improvement was an amphitheater that seated 2,000.[22]

While workers raced to complete the grounds, organizers published a 107-page premium book listing rules, fees, and classes of competition. The fair had to be self-supporting financially, so the fees were explicit. Single tickets for admission were 35 cents; children under 12 years of age paid 20 cents. One- or two-horse vehicles were charged 35 cents, and admission to the grandstand after noon cost 25 cents. Even vehicles running for passenger hire were not exempt; the fee for a two-horse hack was $3, $4 for a two- or four-horse express wagon, and $5 for a four-horse omnibus. Stock entered in the competition was not charged, but each exhibitor had to buy a season ticket for $1.50, and the price of stalls started at 50 cents for an open cattle stall and went to $2 for a box stall. Feed was available at additional cost; straw for bedding was free.[23]

The revenue was necessary to pay the long list of premiums, which ranged from $10 for first prize in some categories to 50 cents for others. The livestock list alone included 66 classes of horses and mules, 50 classes of cattle, 21 classes of sheep, and 26 classes of swine. There also were 54 categories of poultry, 66 categories of farm and garden products, and 164 categories of farm implements. There were competitions for fruit, merchants' displays, art, handiwork, flowers, and food. In all, there were 933 classes to judge everything from pickled picca-lilli to the "oddest curiosity."[24]

Like all fairs in the late 19th century, the main attraction in the program was the Speed Department, better known as horse racing. On Tuesday, the first day of the fair, the schedule included three races: a trotting stake, a trotting stake limited to Oklahoma Territory two-year-olds, and a half-mile dash "open to the world." On the second day, the trotters were joined by a saddle pony race, limited to horses under 14 hands, at a distance of one-quarter mile. The trotting races dominated the final three days, but a little variety was added with a slow mule race and a mile dash for Indian riders. For the five days running, the purse totalled $5,800.[25]

Like every fair in Oklahoma City since 1892, the weather was the first barometer of success. On opening day, one reporter's first paragraph set the tone: "Nature, herself, seemed in harmony with the occasion and Old

Sol as he showed his honest, ruddy face over the fringe of timber which lines the banks of the North Canadian, seemed to be possessed with more than his usual warmth and geniality." The fair itself was not so fortunate. Many of the exhibits were still not on display by noon of the first day, and the horse races were described as "very tedious." The horses were slow in coming on the track, and by 4:30 only one race had been completed.[26]

By the third day the management had matters under control, with one headline declaring it "A Fair that is Surpassing the most Sanguine Expectations in Exhibits and in Attendance." Crowds were estimated at 10,000, with more than 3,000 in the grandstand alone. Beginning a long tradition, each day was dedicated to a specific group, with the third day given to "Farmers Day." Others included "Ex-Confederates Day" and "Peoples Party Day." Although not the beneficiary of a day named after them, the Democrats of the territory convened a rally in Oklahoma City during the fair and flooded the crowds with party loyals from surrounding counties, including one delegation of "160 Normanites." To encourage local

The Oklahoma City Orchestra, seen here in 1905, probably played at territorial fairs (Courtesy Oklahoma Historical Society).

attendance, all merchants in Oklahoma City closed their shops from noon to six on the fourth day of the fair.[27]

Publicly, the first territorial fair was pronounced a success. According to one newspaper account, revenues were enough to pay expenses and leave a surplus to apply on the debts of the association. But the reporter found a greater measure of success:

> The fair has advertised Oklahoma City as it was never advertised before. It has brought hundreds of strangers here from the states, and thousands from parts of the territory that lie outside the radius of our retail trade. They will come again to buy or send in orders. The men who were attracted here from the states will come again

Large crowds gathered along Main Street of Oklahoma City in September of 1899 to view one of many parades held during the second annual Street Fair (Courtesy Oklahoma Historical Society).

Frederick Barde, a correspondent for the Kansas City Star, took this photograph of an exhibitor at the 1st Annual Street Fair in October of 1898 (Courtesy Oklahoma Historical Society).

to buy farms and homes and to engage in other pursuits. They will all bring money with them and in most instances a shiftless farmer will give way to make room for more enterprising ones. All these benefits will result from the First Great Territorial Fair.

The reporter was partially right; the fair would have an impact on economic development, but not for another five years. What he could not predict was drouth, depression, and demoralization, an economic black cloud that soon settled over the town once blessed with the spirit of '89.[28]

Beginning another tradition that would repeat itself well into the 20th century, the downside of the boom and bust cycle was felt first by farmers and ranchers. The low agricultural prices that had driven so many land-seekers to the new Eldorado persisted, held down by over production and weak markets overseas. Then came the drouth, which hit the Southern Plains in 1891 and continued unabated for five lean years. As early as the winter of 1890, federal aid was rushed to Oklahoma Territory where farm families were dying from hunger. Compounding the problem was the fact that titles to land claimed under the Homestead Laws were not granted to owners for five years, which meant the land could not be used as collateral for loans.[29]

The economic boom was fueled by real estate development, pushed forward by businessmen such as Anton Classen, right (Courtesy Oklahoma Metropolitan Library).

In Oklahoma City the effects of the farming depression were initially cushioned by the subsequent land openings of 1890, 1891, and 1892, which pulled people and their money to and through the town as they chased the dream of free land. The construction of the Choctaw Railway east and west through town in 1890 added to the illusion of prosperity, encouraging investments made in hopes of continued expansion. Typically, that illusion was shattered by a single event—a financial crisis in New York City in 1893—and there was no economic foundation to fall back on. In Oklahoma City, all three banks failed, and the young town's first full-scale depression set in.[30]

The economic crisis affected the territorial fair. In the fall of 1893, only a few short months after the worst was realized, the fair board cancelled the exhibits and

competitions and concentrated only on horse races. Lady luck, like the economy, had also turned sour. A cold north wind greeted the horsemen and kept crowds away. As one reporter commented, the "grandstand was a barren waste."[31]

The next year, despite deepening economic problems, the fair board bravely fought back. Again led by C.G. Jones as president and Ed Overholser as secretary, they added new features such as a ferris wheel, premiums for the best township and county exhibits, and a grand stock parade to open the four-day fair. Even the standard displays by merchants had a new attraction, a booth featuring the wines of Edward B. Fairchild, who had planted 20 acres in grapes and built a wine processing vault northeast of town.

Even wine and races, however, could not overcome drouth and depression. The exhibits were, in the words of one normally effusive reporter, "below the expectations of some." Although the financial returns have been lost to history, the results can be deduced from one fact—the 1894 fair was the final act of the territorial fair association.[32]

For the next three years, Oklahoma City did not host a fair or even a horse race. Exhibits were sent to the Guthrie fair in 1896, and a county exhibit was sent to the fair in Wichita, Kansas, in 1897. The only fall celebration was the first appearance of the Ringling Brothers Circus, which drew more than 15,000 people to Oklahoma City in September of 1897. The old fair, the old spirit of optimism and hope, had been shattered, the victim of dry winds and depression followed by financial failure and population decline.[33]

By 1896 Oklahoma City's population had wilted from the original 10,000 to little more than 4,000 people. Meanwhile, less than 30 miles to the north, Guthrie had sustained its original size of 10,000 on the shoulders of corporate investment and political patronage. It was the territorial capital, a Republican town with a healthy economy and rapidly changing skyline pierced by two- and three-story brick buildings that seemed to point the way to a bright and hopeful future. In 1897, as if to acknowledge that supremacy, the business leaders of Guthrie incorporated the first "Oklahoma State Fair."

Led by merchant king William H. Coyle and territorial journalist Frank Greer, Guthrie boosters raised $5,000 and pledged to "conduct a Territorial and State agricultural fair to promote and encourage agriculture, horticulture, mechanics, manufacturing, stock raising and general domestic industry." The six-day fair was conducted October 11-16, 1897, with exhibits and large crowds from Oklahoma City.[34]

Although down, the leadership of Oklahoma City had not surrendered their battle flag. Throughout the lean years from 1892 to 1896, a few city fathers, such as C.G. Jones and Henry Overholser, held fast and continued investing in their community. Their faith was rewarded. The rains finally returned to the Southern Plains in 1896, and farmers began harvesting good crops in 1897 just as commodity prices on the international markets were going up. Wheat doubled in price, and cotton soared to 11 cents a pound. And with the ability to take out mortgages on their homesteaded lands, farmers invested in improvements, machinery, houses, and even luxury goods—all purchased from merchants in town.[35]

An improving economy, like one that is falling, usually is marked and remembered by one single event. In

Early fairs included the hot new sport, football. This was the 1895 team representing Oklahoma High School (Courtesy Oklahoma Publishing Company).

Opposite page: Delmar Gardens was developed in 1903 near the present location of Farmer's Market. It absorbed much of the enthusiasm for fairs in Oklahoma City by providing a constant source of entertainment and special features (Courtesy Oklahoma Historical Society).

Belle Isle Amusement Park, opened in 1907 at the end of the Classen Boulevard streetcar line, was another source of entertainment for the rapidly expanding population of Oklahoma City as the territorial era came to an end (Courtesy Oklahoma County Metropolitan Library).

1897 that symbol of a new age of prosperity ironically followed the first Guthrie State Fair by only a week—the arrival of the first locomotive on the new tracks of the St. Louis and San Francisco Railroad, better known as the Frisco.

Since 1890 the westward extension of the Frisco line out of Sapulpa had been a sought-after prize, the winning city gaining direct trade access to the heart of the Indian Territory and the distant markets of Kansas City and St. Louis. Although Guthrie seemed to have an advantage, being closer and on a direct east-west line from Sapulpa, Jones and Overholser relentlessly courted Frisco officials with promises of cash bonuses, assistance with right-of-way acquisition, and the lure of an agricultural area that was richer than the land around Guthrie. The courtship worked, and on October 21, 1897, the first train pulled into Oklahoma City.[36]

The long-term economic advantages of the new trunk line were undisputed, but Angelo Scott, an '89er and early-day journalist, recognized a greater impact of the Frisco's arrival: "Most of all there was a new psychology, or shall I say the returning of the old psychology. The old hope, the old expectation, and the old spirit came back, and with redoubled strength." Optimism, the keystone of the free enterprise system, was back, unleashing an economic expansion that would make Oklahoma City the fastest-growing city in the entire nation from 1900 to 1910 and end with the removal of the state capital from Guthrie.[37]

The resurrected faith in the future was expressed, as usual, in a fair. In October of 1898, after an unprecedented year of economic improvement, W.W. Storm organized a street fair that was a combination of carnival, trade show, and agricultural exhibit. Storm, a lumber yard owner who later would achieve fame as the president of the city's first streetcar company, raised $3,240 and built a continuous line of canvas-covered booths down the middle of three streets. Foreshadowing the modern Arts Festival by almost a century, these tent booths stretched 500 feet along Grand Avenue, 700 feet along Main Street, and 520 feet north along Broadway. Storm rented the 144 booths to exhibitors and vendors, the price determined by location.[38]

By one estimate, there were 1,200 separate exhibits, with more than half being of farm products, stock, and animals. One of the most remarkable displays was assembled from the produce of one farm owned by Henry Sheplor. Included were "six varieties of corn, three of wheat, potatoes, two of sweet potatoes, one bale of cotton, cotton on the stalk, ginned cotton, two of broom corn, two of kaffir corn, sugar cane, two of oats, pumpkins, 35 barrels of wood, three of apples, two of cucumbers, grapes, five kinds of beans, castor beans, two varieties of watermelon, pie melons, citrons, tomatoes, onions, cabbage, turnips, horse radish, walnuts, persimmons, eggs, acorns, vines, China berries, pomegranate, gourds, butter, canned peaches, canned plums, molasses, grasses, mistletoe, and hay." In all, Sheplor had a total of 113 products from his farm, not including livestock.[39]

Entertainment was a daily staple of the street fair. Grandstands were built at the primary intersections for

The Boardwalk at Delmar Gardens, with nickelodeons in the arcade and a simple menu that offered a dip of ice cream for 5 cents, an ice cream soda for 10 cents, and beer or lemonade for 15 cents. Delmar Gardens offered many of the features of the future fair, such as games, amusement rides, special food, musical events, and the latest in technology, including automobile races with drivers such as the famed Barney Oldfield (Courtesy Oklahoma County Metropolitan Library).

bands, speakers, and entertainers. There were day parades, one with 1,000 school children and another with floats representing various businesses. Night parades were sponsored by groups such as the Shriners, who paid for a "Niagara of fireworks that has never been equaled in the territory." The six-day free fair reportedly drew about 25,000 people, but the organizers still took a loss, leaving, as one reporter noted, "a deficit to be charged to general municipal and territorial good."[40]

Despite the loss, "Oklahoma City's Second Annual Street Fair" was organized and set for September 18-23, 1899. The line of booths was doubled, spanning 3,000 feet along the streets of the central business district, while a greater number of entertainment events were added to the program. In addition to the typical agricultural and trade displays, the organizers offered daily parades, five brass bands, a demonstration of "Edison's talking pictures," a reenactment of Admiral Dewey's famous entrance into Manila Bay, a parachute jump, and a Creek-Choctaw stickball game.

The strategy worked. As one reporter proudly noted, "all roads brought in special trains and each special consisted of ten to fourteen cars, each packed like a sardine can, with pleasure seekers. Every farm wagon within thirty or forty miles was brought into requisition and came into the city filled with boys and girls and older people who came to see the big exhibit and see the sights." Over the six days, more than 50,000 people visited the fair.[41]

For reasons not recorded, the street fair was not repeated after 1899. Most likely, the mild recession of 1900 discouraged organizers, who also would have had to consider the obstacles of congestion, local politics, and an aggressive campaign to pave all downtown streets with brick. Whatever the causes, the only notice of a fair in 1900 was a reference to the Shawnee Fair, while the fall of 1901 was broken only by a short run of the circus.

In 1902 the torch was picked up by Charles Colcord and his partner, Frank Shelley, who organized the "Oklahoma City Fair and Race Meeting." Colcord was another of Oklahoma City's founding fathers, a native of Kentucky who had been a cowboy when he made the run of '89. He became the city's first chief of police, then the first county sheriff in 1890. He left Oklahoma City in 1893 only to return during the boom in 1899. He bought several lots downtown, where he ultimately built the Colcord Building, and several quarter sections west of town, where he and his associates platted numerous housing additions. In 1902, to draw attention to their developments, they created Colcord Park next to

Wheeler Park was the first permanent park in Oklahoma City (Courtesy Oklahoma Historical Society).

The Overholser Mansion, built north of the city at 15th and Hudson in 1902, as seen from the Santa Fe tracks (Courtesy Oklahoma Historical Society).

the Frisco tracks and organized a new fair.[42]

With the stated ambition of a fair that "would not rank second to the Texas State Fair," Shelley built a race track, a midway, and a triumphal arch. He included a few agricultural displays, but most of the grounds were dedicated to entertainment and the races. Reporters noted a shooting gallery, a merry-go-round, and shows that featured performers such as Captain W.F. Drohan, the "oldest Indian scout and mountaineer and a surviving companion of Kit Carson, Jim Beckwith and Tom Dawson." Geronimo, still a prisoner at Fort Sill, was on hand to sign autographed photographs of himself.[43]

The 1902 fair was actually little more than a horse race with side attractions. It would be pretty much the same for the next four years. During the fall of 1903 a Wild West Show drew attention with performers Cole Younger and Frank James. In 1905 a five-day fall carnival was staged near the new courthouse on Main Street, with proceeds going to the Humane Society and the Oklahoma City Rescue Home. All seemed like halfhearted attempts when compared to the territorial fairs of 1892 and 1894 and the street fairs of 1898 and 1899.[44]

Despite numerous attempts, there was no permanent fair organization in Oklahoma City as the territorial era drew to a close. But there were assets to draw upon. First, the seeds of economic prosperity had been sown and nurtured, and by 1906 the economic engines of the community were well tuned and racing ahead. Just as importantly, there was a group of civic leaders who had seen the impact of fairs and felt the potential of bringing thousands of people to the city for fun, education, and profit. And what a well of experience they had to draw from. Learning from both successes and failures, good ideas and bad, they would soon unite behind the boldest effort yet—the first State Fair of Oklahoma.

Chapter Two
A FAIR IS BORN

To the people of Oklahoma Territory, 1907 would be a turning point. Behind were the lean years of opening a new land; ahead was Statehood and all it symbolized—self government, territorial unification, and, most importantly, new opportunities. It was fitting, therefore, that 1907 was also the beginning of a new institution that would play a large role in that future—the State Fair of Oklahoma.

That a fair was even attempted in 1907 is remarkable given the failures of the territorial years. Beginning with the initial attempt in 1892 and ending with a half-hearted effort in 1902, it seemed that Oklahoma City could not generate enough resources, enough enthusiasm, to start a fair and sustain it at a level necessary for success. That legacy of failure would momentarily be forgotten in 1907, swept aside in the wake of an unprecedented economic boom that would provide the resources, generate the will, and unite the city's leadership behind one grand cause.

The spark that ignited the economic explosion began on the farms and ranches of the budding state. From 1900 to 1910 the number of farms in the twin territories almost doubled, from 108,000 to 190,000, while the wealth represented by those farms increased even more. In 1900 the value of all land, buildings, implements, and

Opposite page: The main entrance was a busy place once the fair opened. A circular drive provided access for automobiles, while the streetcar line later terminated just to the right (Courtesy Oklahoma Historical Society).

Oklahoma City's population explosion from 10,000 to 64,000 in only ten years was built largely on the prosperity of farmers and ranchers. During that decade, Oklahoma City became a major center for food and fiber processing, as seen in this old postcard of the Oklahoma Cotton Compress Company (Courtesy Oklahoma Historical Society).

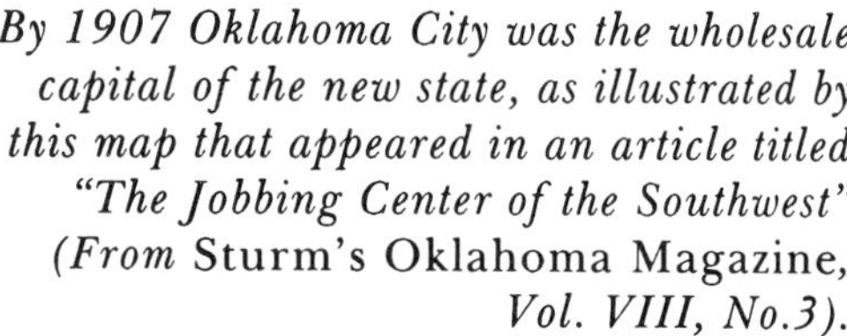

By 1907 Oklahoma City was the wholesale capital of the new state, as illustrated by this map that appeared in an article titled "The Jobbing Center of the Southwest" (From Sturm's Oklahoma Magazine, Vol. VIII, No.3).

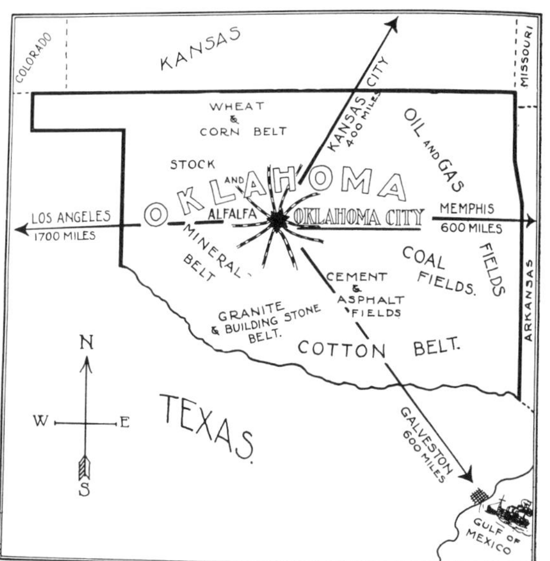

Oklahoma City's Great Wholesale Territory

livestock on territorial farms had been estimated at $277 million; three years after Statehood, the combined value was $918 million.[1]

Leading the way to this new promised land was "King Cotton," a crop particularly well suited to Oklahoma at the turn of the century. The soil was rich and new, unlike the lands of the Old South where repeated plantings had sapped nutrients and destroyed fertility. A poor farmer in Oklahoma, with no more than a mule, a plow, and 40 acres, could plant cotton and expect a return of $18 an acre, compared to $11 an acre for wheat and only $8 an acre for corn.[2]

And plant they did. In 1900 the farmers of the twin territories planted 682,743 acres to cotton; by 1910 the total had increased to 1,976,935 acres. Surprisingly, increased production seemed to have no effect on the price at the gin. From 1898 to 1907 cotton almost doubled, from 6 cents to 11 cents a pound, convincing farmers that good times had finally arrived. With cash in their pockets, they invested in more land, bought new implements, purchased new supplies, and indulged in what had formerly been luxuries such as store-bought dresses, fancy buggies, and faster horses. The circulation of wealth—generated by the bounty of the land— quickly found its way to urban communities, especially Oklahoma City.[3]

By 1907 Oklahoma City was the undisputed economic center of the future state, a stature achieved largely on the shoulders of railroad connections. Of the 6,000 miles of track in the new state, all but 700 were operated by the big four—the Santa Fe, the Frisco, the Rock Island, and the KATY—each of which maintained freight and passenger division headquarters in Oklahoma City. Steel rails linked the city with the best farming counties to the north and southwest and provided easy access to the newly discovered oil fields to the northeast, a factor that would grow even more important as each new discovery made the Mid-Continent Oil Field one of the largest in the nation.[4]

With access to raw materials and growing markets, Oklahoma City attracted industrial investors. From 36 companies in 1900, the industrial base expanded to 89 firms in 1905 and 171 in 1910. Among the new facilities were 44 printing plants, 17 bakeries, 8 lumber mills, 7 flour and grist mills, 8 pharmaceutical companies, and 6 tobacco firms. In one decade, the value of production from Oklahoma City industry rose from $1.2 million to $7.8 million, while wages paid to workers soared from $101,000 to $973,000 a year.[5]

While politics made Guthrie the first state capital, transportation, location, and industrial output made Oklahoma City the wholesale capital of the new state. In

1907 alone, 145 jobbing houses employed 1,300 people in a variety of lines from coffins and confectionery to dental supplies and tinware. According to one annual report, Oklahoma City wholesalers in 1907 sold more than $28 million in products to merchants in all corners of the state.[6]

A large part of those sales were made in Oklahoma City, which was served by 350 retail establishments with a total volume of $19 million at the time of Statehood. Crowding the downtown business district were 109 groceries, 21 meat markets, 21 lumber yards, 18 bakers, 17 tailors, 8 flour and feed stores, 19 drug stores, ten cigar stands, and a host of other sole proprietorships. There also were four national and three state banks; a streetcar system that would soon expand to surrounding towns; and a forest of new residential additions that had already pushed the boundaries of the city north to 16th Street,

Farmers, such as this family on the eastern edge of Oklahoma County, were a primary target of the Oklahoma City Chamber of Commerce, which wanted both their produce and their business (Courtesy Oklahoma Historical Society).

west to Western Avenue, and east to Stiles Circle.[7]

Feeding this economic frenzy was population growth, which not only affected the immediate area of Oklahoma City but also the larger trade territory. From 1890, when the combined population of Oklahoma and Indian territories was 258,657, immigration boosted the count to 790,391 in 1900, followed by an even quicker pace to 1,414,177 by 1907. In Oklahoma City the expansion was even more dramatic, climbing from 4,151 in 1890 to 10,037 in 1900 and 32,452 in 1907. Within just three more years the city's population would double again, to 64,205. Each new resident meant more customers, more capital, more labor, more talent, more fuel for the economic engines of the new state.[8]

Faced with the challenges as well as the opportunities of this runaway economy, the business leaders of Oklahoma City reorganized the old Commercial Club, which by 1902 was described as "long on billiard and pool tables and short on needed office equipment." The new organization was named the Oklahoma City Chamber of Commerce and had a constitution and bylaws modeled after those of the Kansas City Commercial Club. For the first time, corporate and business memberships were accepted along with individuals, and membership grew quickly to 250.[9]

The more aggressive Chamber launched a wide range of traditional initiatives such as farm roads and bridges, the expansion of rail connections, and the promotion of Oklahoma City as an "Eden of Opportunity." Chamber leaders entered the public debate over municipal policies and joined the agitation for public improvements. The

The Oklahoma City Chamber of Commerce, reorganized in 1902, included several men instrumental in the organization of the first State Fair of Oklahoma, including C.G. "Gristmill Jones" (standing, far right), Charles Colcord (standing, fourth from right), and Anton Classen (seated, far right) (Courtesy Oklahoma Historical Society).

staff assumed the duties of the wholesalers' Traffic Bureau, started a Credit Bureau, coordinated the Real Estate Exchange, and increased industrial recruitment.[10]

As the Chamber's mission expanded, it was more important than ever to support projects that could satisfy the needs of a growing and diversified constituency. One of the most effective new tools of accomplishing that mission was the recruitment of conventions. In February of 1903 the Chamber hosted the annual convention of the Oklahoma Livestock Association, beginning a string of successes that would establish the city's reputation as a "convention town." When the Farmers National Congress reported in January of 1907 that it wanted to hold its annual convention in Oklahoma City, the Chamber was ready to listen.[11]

To American farmers, the annual meeting of the National Farmers Congress represented political strength and unity; to the businessmen of Oklahoma City, it meant a financial windfall in convention dollars. There was only one condition requested by the farmers' group. They wanted a fall fair so their delegates could see the achievements of agriculture in the new state. The Chamber leadership, hungry for the convention, agreed to see what could be done.[12]

The commitment came easily, for there had been strong sentiment for a fair as early as 1892. Many of the same men who had put their energies and resources behind that first grand attempt were still in town, still in key positions at the Chamber of Commerce. Moreover, a fair would be good for the community, as pointed out by Frederick S. Barde, a perceptive writer at the turn of the century.

In an article titled "What a State Fair Means to Okla-

Oklahoma City had already become a sprawling suburban city in 1907 when the Chamber of Commerce sought a site for the first state fair. This photograph was taken in 1907 at what became N.W. 39th and Classen Boulevard, where the streetcar tracks split, one branch going north to Belle Isle Power Plant, the other branch curving west toward the small town of Bethany (Courtesy Oklahoma County Metropolitan Library).

homa," Barde listed the reasons why economic boosters should support a fair. "Where else," he asked, "could people from other states find in a single place samples of everything made in Oklahoma?" At the same time, he added, "this visible evidence of Oklahoma's greatness would inevitably arouse a greater and deeper state pride among citizens...founded upon facts, not prejudices ...and felt by the boys and girls growing into manhood and womanhood."[13]

Barde also recognized the impact a fair could have on reducing the sectional divisions of the new state, which had been formed by combining the old Oklahoma Territory with the Indian Territory. "This geographical line between the two former territories," he wrote, "grew from an imaginary nothing to an intangible something. The fact that one man prefers biscuits and ham gravy to lightbread and flour gravy is no good reason why he should be made an outcast." "How," he asked, "could a better way be devised for the removal of sectional prejudice, socially and politically, than by bringing the people of the state together once a year in the contemplation and enjoyment of those things which all have labored to produce?" To the business leaders of Oklahoma City, this vision of a unified state meant a more efficient economic machine—and efficiency meant growth.[14]

Another writer, W.M. Sturgis, expressed more traditional reasons to support a fair. "A State Fair," he wrote, "is the best and the most profitable investment possible for the betterment and advancement of the whole people.... It promotes industry, fosters enterprise, generates competition, gives fresh motive to farmers and stock raisers and encourages agriculture." Appealing to civic pride, he pointed out that "every state in the nation has its State Fair and Oklahoma, not to be outdone by her older sisters, should have the greatest show of the year."[15]

Convinced of the potential, a coalition of business leaders gathered in the offices of the Chamber of Commerce on January 18, 1907, to explore the possibilities of "organizing a state fair association." Presiding was C.G. "Gristmill" Jones, the same man who had led the effort to create territorial fairs in 1892 and 1894. Alexander W. McKeand, managing director of the Oklahoma City Chamber of Commerce, served as acting secretary. Among the crowd gathered that Friday afternoon were Anton Classen, former president of the Chamber, Weston Atwood, current vice-president of the Chamber, and a who's who of the citizens of Oklahoma City.[16]

Charles Colcord was an '89er who made a fortune in real estate and oil. When the Chamber of Commerce initiated plans for a fair, he exerted all of his powers of influence to build the new institution on the west side of town, near his many real estate ventures (Courtesy Oklahoma Historical Society).

The west side location had many advantages, as seen in this photograph looking west on Main Street. The new Oklahoma County Courthouse (with the tower on the left) was located there, and an existing double streetcar line ran directly to Delmar Gardens, one of the sites proposed for the first state fair (Courtesy Oklahoma Historical Society).

Early fund-raising efforts for the fair were based on stock subscriptions (State Fair Archives).

The meeting started with general discussion of location, organization, and management, but quickly turned to financial needs. From lessons learned during the territorial years, it was clear that capitalization was critical if the effort was to survive the initial investments in buildings and attractions. On the motion of Atwood, the group voted to charter the State Fair Association of Oklahoma with a capital stock of $100,000, divided into 10,000 shares valued at $10 each.[17]

Then the discussion returned to the issue of location. Some believed it would be impossible to hold the initial fair on land owned by the association due to the short preparation time. They suggested an operating agreement be made with the owners of Delmar Gardens, a popular amusement park located a mile west of town. Others insisted on buying land as soon as possible so permanent improvements could begin. Faced with this divisive issue, Chairman Jones named one committee to seek a charter and another to pursue the options for location.[18]

On January 21, 1907, the Secretary of the Territory issued a charter of incorporation to the State Fair Association of Oklahoma. Four days later, at the association's second meeting, Jones named two more committees, one to draft a constitution and bylaws, the other to solicit subscribers for the stock. After several emotional speeches, the crowd demanded that the stock books be opened immediately so they could pledge their support. C.G. Jones and Charles Colcord each took 50 shares, followed by 12 others who subscribed to ten shares. Subscription blanks were passed around, and salesmen hit the streets. Within the month, pledges rose to $24,990.[19]

The issue of location was not so easily settled. By the second meeting it was clear there were two choices, one east and one west, each with a faithful following among the charter stockholders. The west side faction was led by Charles Colcord and his real estate partner, Frank H. Shelley. As early as 1900 Colcord had bought large tracts of farm land west of the city, where he platted several housing additions and opened Delmar Gardens under the management of the Sinoupolo brothers. They were joined in the crusade by I.M. Putnam, another capitalist who had invested heavily in land to the west. Initially, the west side faction suggested a contract arrangement with Delmar Gardens, then turned their attention to the county poor farm, which was located another mile to the west.[20]

The east side advocates were led by three farmers, Vincent L. Bath, Dr. Francis M. Jordan, and A.H. Keller, all of whom owned land adjacent to a quarter section of school land almost two miles east of the city on the banks of the North Canadian River. Bath, who had come to

Oklahoma City in 1900, had already platted several
housing additions on his land and saw the fair as a means
of attracting buyers. He held the lease on the school
land, but was willing to give it up for the fair. Dr. Jordan
was a respected '89er who had homesteaded the quarter
section immediately west of the school land. He had
prospered, with extensive orchards, an irrigation system,
and a dairy. Their case for the east side location was
strengthened when it was reported on February 12 that
most of the stock subscriptions had been taken by farmers east of the city.[21]

While the two factions lined up votes, the organizing
committee made progress. On February 25 the directors
adopted an amended version of the constitution and
bylaws of the Dallas State Fair, then voted to send Jones
and Shelley to Dallas to investigate facilities and management systems. It was a natural choice for emulation.[22]

One of the first buildings constructed on the new fairgrounds was the Entrance Building, a frame structure with two offices on either side of the arcaded entry (Courtesy Oklahoma County Metropolitan Library).

C.P. Sites, the first staff member of the State Fair (State Fair Archives).

The State Fair of Texas was the largest annual exposition in the Southwest, tracing its roots to 1858. Like the territorial efforts in Oklahoma City, however, the early years of the Texas fair had been plagued by problems. It was interrupted by the Civil War, then resurrected in 1872. After fourteen years of experimentation and change, it was chartered by the state and the first structures were built at Fair Park. Still, the problems had continued. In 1886, the first year at the new location, there was a $140,000 deficit, followed by fires in 1888 and 1891. Not until 1892 did the fair board declare a small surplus. When Jones and Shelley crossed the Red River for a visit, their counterparts in Dallas were in the second year of a building campaign.[23]

By March the location battle was back on the table. Although President Jones probably favored the east side location, which was in the direction of his many investments, he never publicly stated a preference in an attempt to keep all factions united behind the effort. At one public meeting, I.M. Putnam made a motion to appoint a new committee on location. Jones ruled him out of order. Putnam countered with another motion to reconsider an earlier vote on the east side location. Again, Jones ruled him out of order. On March 29, Jones relented and appointed a new committee of five to report on location. To the committee he appointed Anton Classen, Oscar G. Lee, Dr. Jordan, Vincent Bath, and Charles F. Colcord. Not surprisingly, the recommendation of the committee was three votes for the west side location, two for the east side school land. The final decision, suggested Classen, should be put before the stockholders.[24]

On April 12, 1907, the full board of directors, officers, and "numerous shareholders" met in the Chamber offices to vote on location. Before the vote, it was decided that any subscribers who were not satisfied with the final decision could withdraw their pledges. After voting to suspend the rules, C.H. Keller moved to locate the fair on the east side, lease the school land, and buy additional land so they could build a one-mile track. The motion carried.[25]

Two weeks later four of the original directors resigned, including Lee, Colcord, and Shelley, who had been serving as the first paid secretary of the association. Filling their places were Samuel Gloyd, owner of several lumber yards, J.G. Leeper, a businessman, and Henry Overholser, one of Jones's closest business and political allies listed in early city directories as "H.J. Overholser —Capitalist." To replace Shelley, the board hired C.P.

On October 3, 1907, with the first fair only two weeks away, J.A. Ryan purchased five shares of capital stock in the State Fair Association of Oklahoma. The certificate was signed by C.P. Sites, secretary, and C.G. Jones, president (State Fair Archives).

Sites as secretary of the association. His salary was
$2,400 a year.[26]

With the location battle settled, Jones and the board
turned to finances and preparation for the fall fair. On
May 10, Sites reported that subscriptions were down to
$33,190 after the withdrawals had been made, and of that
amount, only $1,800 had actually been paid. Before
work could begin, Jones said, $50,000 had to be pledged.
The challenge worked. Within five days subscriptions
were up to $45,000, and the remaining $5,000 in pledges
were taken by board members.[27]

Jones and Sites prepared themselves for the coming
task by visiting the Wichita State Fair and the Dallas State
Fair again where they met with management and studied
facilities and programs. Back in Oklahoma City, Jones
appointed a Purchasing and Building Committee consisting of himself, Sites, and board member J.G. Leeper, and
rented a room for Sites next door to his office at 11 N.
Robinson.[28]

Together, they completed the lease on the school
land, which cost $140 a year, and hired Edward Coady as
architect. They submitted plans for an Exhibition Building, an Administration Building, an Amphitheater, and
an entrance gate, which the board approved. They also
started work on the premium list and booked the famous
Seymour Military Band for $3,700. When bids for con-
struction exceeded their expectations, they decided to
build the structures with day labor and hired a general
superintendent for $7 a day. Work on the grounds
began Monday, June 30, 1907.[29]

The site was a 160-acre tract that had been planted to
corn and pasture. It was flat, with rich soil, a small lake,
and a grove of trees situated between the Deep Fork River
on the north and the North Canadian River on the south. A
branch of the old Chisholm Trail had crossed the river
only a few hundred feet to the east. Located less than
two miles from the heart of the downtown business district, it was already surrounded by the early signs of
suburban sprawl. Transportation to the site, a critical
concern during the site selection process, was to be provided by a special switch of the KATY railroad that would
terminate directly in front of the entrance gate.[30]

Work began immediately on the half-mile track,
scaled back from the ambitious one-mile distance when
additional land proved too expensive. The surface was
scraped and "made dirt" was backfilled to give the track a
"springy" action. Meanwhile, work started on the grandstand, designed to seat 5,000 people. Facing the east, it
was a wooden structure elevated on hardwood beams and
covered by a slightly sloped shed roof.[31]

Between the grandstand and the ornate entrance gate,

*This complimentary press pass admitted
one reporter to the first State Fair of
Oklahoma (State Fair Archives).*

crews started work on the Exposition Building, a two-story frame structure 175 feet wide by 225 feet long designed for the display of commercial and manufacturing exhibits. Next door to the south was the Music Hall, measuring 80 by 150 feet, with a grand stage and comfortable seats inside. Completing the grounds was a Poultry House, 48 feet wide by 72 feet long; several barns for cattle, horses, sheep, and hogs; the midway, consisting of several frame buildings 270 feet long and 24 feet wide; and a fence around the entire complex. By the time construction workers were finished, one reporter estimated that $125,000 had been spent on the grounds.[32]

After a full summer of planning, working, and scrambling to raise money to pay the bills, the first State Fair of Oklahoma was ready for opening day, October 5, 1907. Good weather greeted the day, as more than 10,000 people crowded onto the grounds even before the opening ceremonies began at 10:00 a.m. At the appointed time, C.G. Jones, president of the Fair Association, ascended the podium and helped his eleven-year-old son push the button that signaled the opening of the fair. Peterson's Band broke into a rousing version of "The Fair March," dedicated to President Jones.[33]

Sharing the podium was Governor Charles N. Haskell, who had only recently been elected first governor of the

The first fair included a wide variety of farm produce and plants, including a display on Wheeler Park assembled by the Oklahoma City Parks Department (Courtesy Western History Collections).

new state. Addressing the crowd, he praised the fair and the people who had made it a reality:

> C.G. Jones and other able and public spirited men have given so freely of their time and their money to make this fair a success.... The people of Oklahoma City may feel justly proud not only of the success of the fair, which is assured, but upon the possession of men with the public spirit, the energy and the ability to organize the movement.[34]

At the conclusion of the governor's speech, Helen Renstrom, billed as "the Oklahoma Nightingale," sang "Oklahoma: A Toast" to a standing ovation. The fair was off to a good start.

The program assembled by Jones, Sites, and the board made it abundantly clear that the State Fair of Oklahoma was first and foremost an agricultural fair. "The farmer is to have the front seat at the great fair in all things," wrote one of the fair's most vigorous proponents. "It is to be primarily an exposition of the products of the farm, a place where not only will be found the corn, wheat, cotton, broomcorn, potatoes, alfalfa and all that the Oklahoma farmer grows, but all of his livestock, in fact everything he raises, even to the garden and the orchard."[35]

To lure farmers to enter their crops and livestock, the fair board approved a premium list awarding $30,000 in prizes. The strategy worked. By the opening day of the fair, the horse barn was full and one of the exhibit cattle barns had been converted for horses. Extra stalls

The road in front of the original Exposition Building was lined with midway acts and games (State Fair Archives).

"Oklahoma's Family at the Fair," an editorial cartoon run in Oklahoma City newspapers, emphasized that the State Fair of Oklahoma was "first and foremost an agricultural exposition" (Untitled clipping in the State Fair Archives).

Oklahoma's Family at the Fair

The official "Souvenir Badge of the First State Fair," emblazoned with the seal of the State of Oklahoma (State Fair Archives).

were built in the remaining seven cattle barns. More than 2,500 chickens were entered, including one flock of 300 birds shipped in by a farmer from Wisconsin, making the poultry division the largest in terms of entries.[36]

In the livestock department, 352 exhibitors entered 1,643 animals. The list included 37 exhibitors with 380 cattle, 76 exhibitors with 282 horses, 86 exhibitors with 335 swine, 16 exhibitors with 103 sheep, and 9 exhibitors with 36 dairy cattle. The program also listed premiums for the Tri-State Bench Show with 20 exhibitors showing 76 dogs, including James Meyers of Tuttle who entered a pack of 20 English foxhounds that had caught and killed 50 wolves and coyotes within the previous twelve months.[37]

In the general farm section, more than 125 farmers entered displays, led by Vincent "Farmer" Anderson who boasted that he had produced or manufactured more than 300 different items on his farm north of Oklahoma City. All were on display. Oklahoma A&M, the agricultural college in Stillwater, set up a farm demonstration booth and advertised it would be giving away samples of homemade biscuits and honey.[38]

To fan the flames of sectional competition, the premium list included $500 for the best three exhibits assembled by counties. Answering the call were Dewey, Pittsburg, Canadian, Oklahoma, and Woods counties. Notable was the Woods County display that included a scale model of the entire town in miniature showing the courthouse, college, business district, and a scattering of homes. There was a "hand" of tobacco grown at Watonga, a stalk of corn 14 feet high, and "sweet potatoes too large for the oven and beef pan."[39]

In addition to the variety of farm products, county displays included samples of natural resources and manufactured items. Pittsburg County exhibitors showcased a block of coal taken from a Krebs mine that weighed 6,700 pounds and measured four by six feet. From McAlester came decorative bricks made from a mixture of gypsum and wood pulp. The Woods County display featured a box of crystal sand used in high grade finishing work. Previously imported into the territory, the crystal was discovered three miles west of Alva in a vein more than three feet thick.[40]

Next to the county displays on the first floor of the Exhibits Building were the booths of companies and traveling salesmen, where merchandise ranged from dry goods and hosiery to stoves and sewing machines. On the last day of the fair many of the merchants and salesmen donated their stock for a raffle, with all proceeds donated to the needy widows and orphans of the new state.[41]

The Art Department was on the second floor of the Exhibit Building. In the fine arts division, the featured

"Race Day" at the first fair (State Fair Archives).

display was the art of "priest-artist" Father Gerrer, who won four blue ribbons with his paintings "An Oklahoma Sunset," two local portraits, and his famous portrait of Pope Pius X, a copy of which was hanging in the Vatican. There also were tapestries, laces, and quilts, including one diamond pattern quilt of more than 1,700 pieces.[42]

A popular area was the Curio and Antique Department. Items on display included a piece of hand-hammered native copper that had belonged to Osage Chief Pawhuska, a beaded hunting bag belonging to Chief Jim Bob of the Delaware, and a collection of "Aztec pottery" found in a mound in the eastern part of the state. The pieces, undoubtedly from mounds later identified as part of the Spiroan Culture, included two vases, a figurine, and a bracelet in the form of a coiled snake.[43]

Indian culture, which still flourished through much of Oklahoma, was a growing curiosity with the general public, both in and out of state. At the fair, an Indian village was recreated with 48 Cheyenne and a few Pottawatomies and Sioux, led by Cheyenne headmen Cloud Chief, Little Bear, and Buffalo Meat. They set up camp, performed what one reporter called the Sun Dance, and displayed handiwork and weaponry.[44]

A more exotic demonstration was the Igorrote Village, an attraction billed as "twenty-five head hunting, dog-eating wild people from the Phillipines." According to one promotional description, it was a tribal custom "for the Igorrote men to eat dogs, but the women are not allowed to partake of this delicacy, because it is considered to have a quality which nerves up the fighter for

head hunting expeditions." The men, "meagerly clad, wearing a half coconut set jauntily on the backs of their heads," staged war and ceremonial dances, sang tribal songs, and demonstrated skills such as spear throwing.[45]

Curiosity mixed with a desire to be entertained were strong emotions that pulled people to the first State Fair of Oklahoma. The main attraction was Sooners' Trail, forerunner of the midway. A feature at the freak show was the "Australian Snake Girl" who handled Australian rattlesnakes, "winding them with loving tenderness about her neck, kissing and caressing them with impunity." Farther down the trail was Bob Carroll's Carnival company with six vaudeville shows, animal acts, minstrels, and two "electric theaters."[46]

For sheer entertainment, a highlight was the "Carnival of Naples," staged in front of the grandstand. The set included "acres" of realistic buildings, temples, and street scenes of ancient Naples "nestled at the foot of frowning Vesuvius." At 8:00 pm each night, 250 costumed actors, acrobats, high wire artists, and musicians made a grand procession onto the stage where they hailed the Italian king and queen, celebrated with the festival of lanterns, and danced to the "Maypole Balletto." The climax of the performance was the eruption of Mount Vesuvius with sound effects and a "river of living fire." The destruction of the set was followed immediately by a fireworks display that "filled the Heavens with glorious colors."[47]

As popular as the spectacle of fireworks and flying acrobats must have been, the featured attraction of the fair was reserved for a more traditional, more familiar event that appealed to the people of Oklahoma—horse racing. The popularity of speed and wagering was as old as the territory itself. George Catlin, an artist who had visited the Southern Plains in 1834, wrote that as horsemen the Comanches were "not equalled by any other Indians on the continent. Racing horses, it would seem, is a constant and almost incessant exercise, and their principal mode of gambling." Horsemen of the Five Civilized Tribes shared in the passion. Tracks with weekly races were featured at towns such as Parris Prairie, Muskogee, and Checotah, while every local fair and celebration usually included a few races, even if it was just to prove who had the fastest horse in the community.[48]

After the land runs, horse racing quickly became a favored pastime in Oklahoma Territory as well. There were so many tracks and races by 1900 that a circuit with standardized formats had been established with racing cards in Enid, Hennessey, El Reno, Oklahoma City, Guthrie, Perry, and Newkirk. Another organization that same year had scheduled races for Frederick, Mangum, Elk City, Clinton, Canton, Fairview, Cherokee, Watonga, El Reno, Walters, and Wichita Falls, Texas.[49]

In Oklahoma City horse races had been staged only three months after the land run, and were repeated at every possible opportunity. The territorial fairs of 1892 and 1894 had featured races, as had the street fairs of 1898 and 1899. Even when the fairs were cancelled, the races went on. By 1907 there were three permanent tracks in Oklahoma City: at Colcord Park, Delmar Gardens, and Kramer Park.[50]

At the organizational meeting of the State Fair Association, one of the first topics of discussion was a track. Charles Colcord wanted a one-mile track rather than the typical half-mile length. Regardless of the size, everyone assumed the track and grandstand would be central features of any plan no matter where the fair was located or how it was financed. When work began on the school land site, the first dirt turned was for the grandstand and track.[51]

Although Jones and Sites had scheduled several days of racing with trotters, runners, and exhibitions, the highlight of the race meet was the Oklahoma Derby, a one-mile race on October 10, 1907. Sixteen horses were entered, including favorites Miss Affable, a horse that had "run away with everything but her shadow" in Kansas City, and R.Q. Smith, a horse that had already won several races on previous days of the fair meet. A long shot was Buster Jones, known as a good but "capricious colt" that backed into the scratch, or starting position.[52]

On the day of the race, more than 15,000 fans crowded

Governor Charles N. Haskell (center in dark suit coat) addressed the crowd at opening ceremonies for the new State Fair (Courtesy Oklahoma Historical Society).

into and around the grandstand. General admission cost 50 cents, reserved seats 75 cents, and box seats $1.00. Gambling was heavy with legal bookmakers and touts shouting their willingness to do business as post time neared. After several false starts earned a warning that the judges would penalize the "more ambitious jockeys," the starter shouted "go." Buster Jones, with Jimmy Donivan aboard, bolted to the lead, paced the field around both laps, and held on as the horses pounded to the finish line. The surprise winner paid 6 to 1.[53]

Derby Day was only one of several special "days" featured by the fair. The first was Kentucky Day, with a reunion of all Kentuckians in the state. College Day paid honor to higher education, with a speech by Dr. Angelo Scott, president of Oklahoma A&M, and a football game between Oklahoma City's Epworth University and Northwestern Normal School in Alva. Missourians' Day was followed by Oklahoma City Day, with all schools and businesses closed so all could decorate their homes and attend the fair. Friday, October 11, was Old Soldiers' Day, with special programs and parades for veterans of the "Blue and Gray," the Spanish American War, and the Rough Riders Association.[54]

Traveling Men's Day started with a street parade hosted by 2,000 traveling salesmen and concluded with a dramatic performance of Nels Darling in "The Life of a Traveling Man." The final day of the fair was Union Labor Day, featuring speeches by S.O. Dawes of Shawnee, president of the State Farmers' Union, J.L. Langston, secretary of the territorial labor unions, and J. Harvey Lynch, editor of the *Union Messenger*. Each speaker was scheduled in the Music Hall around the daily offerings of concerts by Seymour's Military Band and performers such as Madame Esmathilde, renowned violin virtuoso, and Madame D'Alberti, celebrated dramatic soprano.[55]

On October 16, 1907, after extending the run an extra day for delegates of the National Farmers Congress, the first State Fair of Oklahoma came to a close. In terms of exhibits and entertainment, Jones and Sites had reason to boast. Hundreds of farmers had displayed the bounty of the land, while exhibitors had showcased a wide range of raw materials and finished products for potential investors. And although official figures were not publicized, total attendance likely exceeded 75,000 people, most coming from out of town. Every visitor came to Oklahoma City with money to spend and left with memories of services and products for future use.[56]

Despite the glow of success, there were problems. Revenues were not enough to pay expenses, and most of the premiums went unpaid. Bills for the original construction, completed with promises and optimism, came

The fair was both an opportunity to show the best products of the farm and a chance to visit the big city (Untitled clipping in the State Fair Archives).

due. And the general economy, so healthy when stock subscriptions had been pledged, turned sour after another national banking crisis triggered a financial crash in Oklahoma City. To C.G. Jones, it probably seemed like 1894 again when the territorial fair had failed under the weight of debts and hard times. This time, however, history would not repeat itself.

The biggest building on the new fairgrounds was the Exposition Building, another frame structure painted white. The Entrance Building was located to the left (Courtesy Oklahoma Historical Society).

SAC AND FOX
AGENCY
CHE
AS YE
SOW

INVESTING
IN THE FUTURE

It was a promising beginning. The State Fair of Oklahoma, a combination of education, entertainment, and boosterism, had been pulled together and opened to an enthusiastic public in nine short months. As the fair board quickly discovered, that first step would prove to be the easy part.

Even before the gates of the inaugural fair closed, it was obvious the fair association faced serious financial problems. There was no money in the treasury to pay the premiums, which amounted to more than $3,000, and construction bills were growing daily. Work on the original buildings, it turned out, had been done on credit.

The reasons for the shortfall were not hard to understand. Subscriptions pledged in the early enthusiasm for the new fair went unpaid as businessmen struggled to cope with the aftershocks of the Panic of 1907, a banking crisis that started on Wall Street and plunged the country into a brief but dramatic depression. Then there were the monumental start-up costs of site preparation and construction, an investment that normally would be amortized over many years. In 1907, without adequate capitalization, the entire amount came out of cash on hand.[1]

Opposite page: Displays from the various Indian agencies were featured for several years. Here, the Sac and Fox Agency included handiwork as well as agricultural output (Courtesy Western History Collections).

The OPUBCO Building, constructed in 1909 and 1910, was another building designed by Layton and Smith, the architectural firm that developed the fairgrounds expansion in 1917 (From Sturm's Magazine, *Vol. VIII, No. 1).*

Faced with disabling debts and no way to raise money, the fair board turned to the one man who could marshal the resources and leadership needed to save the fair. That man, soon to be known as the "father of the State Fair," was Henry Overholser.

Overholser's background had prepared him well for the challenge. Born the son of German immigrants in 1846, he was a product of the American frontier, a true believer in the notion that any man could succeed if only he worked hard, applied what he learned, and was willing to pull his weight in the community. His own business career confirmed that belief. He had begun as a merchant in Indiana, then joined the rush to the silver mines of Colorado and the iron ore fields of Minnesota. By 1889, after several years in Kansas, the 43-year-old adventurer decided to take one more chance—the opening of America's last frontier, the Oklahoma country.[2]

Overholser was not the stereotypical land rusher on horseback. Unlike most of his fellow 89ers, he arrived in style aboard the train, bringing with him capital for investment and several boxcars full of prefabricated buildings. He purchased lots at what would become Grand and Robinson and erected six two-story buildings within a few weeks. Subsequent investments in the young community included the three-story Grand Avenue Hotel, the brick United States Courthouse, and the Grand Avenue Opera House, all completed by 1891.[3]

Overholser's enthusiasm for investment in the future was not limited to buildings. He joined C.G. Jones in capitalizing the Grand Canal during the bleak winter of 1889-1890, stopped a bank run during the Panic of 1893, invested in one of the first power utility companies in town, and helped build the first long distance telephone lines to Oklahoma City. Even more importantly, he contributed his influence and capital to the successful recruitment of the Frisco Railroad in 1897, a turning point that started Oklahoma City's unprecedented population growth from 4,000 to 64,000 people in less than 13 years.

Throughout this period of his business career, Overholser was one of the most committed advocates for a fair, a role he assumed quietly and without fanfare. When C.G. Jones organized the territorial fairs of 1892 and 1894, Overholser had been the largest stock subscriber. When funding was needed for the 1907 state fair, Overholser again had been the largest investor. Every time there was a community effort to help build Oklahoma City during its formative years, Overholser was in the ranks if not in the headlines.[4]

With the first State Fair on the brink of bankruptcy, Henry Overholser decided it was time to step out of the shadows. On November 12, 1907, he arranged a $23,000 loan to the fair association, payable at eight percent interest over five years. Two days later, he negotiated another loan for $25,000. Both mortgages were financed

Removal of the state capital from Guthrie to Oklahoma City in 1910 was another boost to local economic development. Here, a billboard touts the benefits of the move (Courtesy Oklahoma Historical Society).

officially in the name of Queen Pirtle, Eldorado, Kansas. In fact, Queen Pirtle was the daughter of Henry Overholser, and the money was his.[5]

With overdue bills paid, Overholser turned his attention to reorganizing management of the fair. At a board meeting on December 24, 1907, he offered to serve as secretary for one year, without salary, provided the fair board employ an assistant secretary. J.T. Sites, who had served as the first secretary, said he did not want to stand in the way of such a generous offer and tendered his resignation. Two days later, Overholser hired his new right-hand man, I.S. Mahan.[6]

Isaac Shepherd Mahan would prove a wise choice. Born the son of a prominent attorney in Lexington, Illinois, he had attended the University of Iowa, earning the degree of Doctor of Dental Surgery, and entered the practice of dentistry. In 1907, at the age of 34, he heeded the call of the West and moved to the nation's fastest growing boom town, Oklahoma City, where he entered the real estate business. Although it is not known how he came to Overholser's attention, it is likely that it was through their mutual devotion to the Republican Party. Whatever the reason, in January of 1908 Mahan started his new career with the State Fair of Oklahoma. His salary was $75 a month.[7]

For the next five years, Overholser and Mahan would dominate the affairs of the State Fair, an arrangement that apparently suited both the strong-willed Overholser and the exhausted fair board. C.G. Jones, the founder who had been reelected president of the board of directors, would be absent much of the next two years, serving in Guthrie as the state senator from Oklahoma County. Dr. Jordan, the vice-president, was elderly and far from assertive. And whereas the first fair had been organized and administered by various board committees, the mention of committee work virtually disappears from the minutes of the board of directors after January of 1908. Although it was never stated so bluntly in the board

This post card featured a night shot of the fairgrounds, made possible by the investment in lights for evening activities (Courtesy Western History Collections).

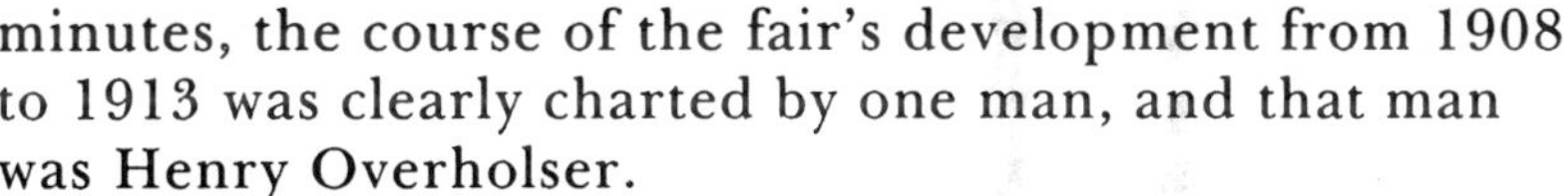

minutes, the course of the fair's development from 1908 to 1913 was clearly charted by one man, and that man was Henry Overholser.

With overdue bills paid and management consolidated in his own hands, Overholser adopted the battle cry of his long business and civic career, "Invest in the future." The most aggressive program, and the one most desperately needed, was in capital improvements. The first fair had been confined to four primary buildings: the Grandstand, the Office Building\Entrance Gate, the Exhibits Building, and the Music Hall. None were particularly impressive. The Exhibit Building had dirt floors, the grounds were prairie grass in spots, and the entire complex looked like a frontier race track surrounded by temporary buildings. If the fair was to grow and prosper, Overholser knew, the number and quality of the facilities had to improve.

The Oklahoma Geological Survey sponsored exhibits in the Minerals Building, as seen in this display in 1909. Such attractions promoted development of resources and lured the attention of outside investors, a critical need on the capital-starved frontier (State Fair Archives).

First, there had to be a better way of getting people to the fair. The fairgrounds was located two miles from the downtown business district, too far to walk comfortably. In 1907 the railroad ran special trains, but the number of passengers carried was far too small, and the two-mile circuit was too short for the efficient operation of steam engines. The solution was a streetcar line. Overholser, who had been instrumental in chartering the first streetcar line in 1902, used his connections to get the company to build a line to the front entrance of the fairgrounds. By the next fair, 30 electric streetcars would run on the line, arriving every five minutes as their drivers circled to a stop in front of the main gate.[8]

Overholser also wanted to pave 8th Street, the primary east-west avenue to the fair. In 1907 the constant flow of buggies and automobiles had stirred up storms of dust on dry days and bogs of mud on wet days. Lot owners along the route, however, were unwilling to vote for a tax assessment to pay for the paving. Overholser, as usual, put up $14,000 of his own money to buy the lots. As he said in a newspaper article, "Someone friendly to the fair had to buy them in order to have the paving proceed."[9]

Pressed by financial need, the board of directors increased the capitalization of stock from $100,000 to

$200,000 in April of 1908 and vowed to raise more money for improvements. That effort, like those the previous year, fell far short of needs. By June of 1908 only $53,958 had actually been paid on all stock subscriptions, including those pledged the previous year. Again, Overholser dipped into his own pockets and loaned another $23,000 to the fair association for improvements during the summer of 1908, bringing his total cash outlay to more than $85,000.[10]

With this new infusion of funds, Overholser and Mahan initiated a long list of new projects. They constructed an Administration Building, which housed the secretary and provided space for a police station, a Western Union telegraph office, an information bureau, and an emergency hospital. They doubled the number of barns and constructed the Poultry Building, 73 feet wide by 140 feet long, big enough to display more than 4,000 fowls.[11]

They added a cement floor to the Exhibit Building and converted the original Music Hall into additional exhibit space. To accommodate larger crowds for music, racing, and vaudeville acts, they built a bandstand outside and enlarged the Grandstand to seat 7,000 people. Overholser and Mahan also invested in beautification of the grounds. They laid more than 140,000 feet of concrete sidewalks and driveways, scattered more than 1,000 bench seats, and installed an outdoor lighting system that turned the nighttime fairgrounds into a "white city." By the time the fair opened on September 21, 1908, there was little resemblance to the first fair.[12]

As Overholser predicted, the investments paid handsome dividends. Total attendance at the 1908 fair was estimated at 100,000, a 25 percent increase over the first fair, and the number of exhibitors increased to 822. More importantly, the bottom line improved. Revenues were $60,000 while expenditures, not counting capital improvements, were $46,000. Even if a $14,000 profit did not pay for improvements or interest on long-term notes, it at least kept the fair from going further into debt just to pay for operations.[13]

Encouraged by success and Overholser's faith in the future, the fair board roared ahead with plans for further expansion. In December of 1908, after Jones resigned as president, the board elected Overholser president and Mahan secretary and gave them a virtual mandate to continue the capital investments and improvements that had been so productive the past year.

Their confidence was not based solely on the abilities of Overholser and Mahan. Just as important was the economic boom raging all about them. The decade-long economic expansion that had been momentarily dis-

I.S. Mahan, trained as a dentist, was Henry Overholser's choice as the first man to run the day-to-day operations of the expanded State Fair of Oklahoma. Mahan would go on to a distinguished career in fair management, including leadership positions in several national associations (Untitled clipping in the State Fair Archives).

rupted in 1907 was back on track and stronger than ever. In the winter of 1908, when Overholser became president of the fair board, Oklahoma City was riding the crest of a three-year population explosion from 32,000 to 64,000. Construction of homes was booming, with new neighborhoods stretching along the streetcar lines east to the Maywood Addition, west to Colcord Heights, and north to Anton Classen's Highland Park Addition, later to be renamed Heritage Hills.[14]

The downtown business district looked like a war zone with recently constructed two- and three-story brick buildings coming down to make way for new and larger buildings such as the Pioneer Telephone Building, Oklahoma City's first skyscraper, and the Huckins, the city's first elegant hotel. Within the year, construction would start on the Skirvin Hotel, the Baum Building, the Colcord Building, the Oklahoma Publishing Company Building, Central High School, the White Temple, and scores of other structures. It was the boom of the century, and the leaders of the State Fair of Oklahoma were ready to plunge ahead, certain that the future would be even bigger and better.[15]

For the next three years improvements on the fairgrounds set a torrid pace, funded by a series of loans starting with a $20,000 note taken by Overholser in the spring of 1909. First came a new Agriculture Building, with 15,000 square feet of space to be used for county exhibits, farm competition, horticultural and floricultural displays, and bee and dairy exhibits. A private firm, the

View from the Grandstand, looking northeast over the track on race day in 1908 (Courtesy of Mrs. W.B. Stephens, Harrah, Oklahoma).

The Lee-Huckins Hotel, opened in 1910 during the building boom, was one of the many local businesses affected by the fair every fall (From Sturm's Oklahoma Magazine, *Vol. IX, No. 6).*

Oklahoma Fair Park Amusement Company, joined in the expansion and paid for three new rides: the Carousel, the Figure "8", and the Canals of Venice, a $12,000 indoor attraction that carried patrons by boat along a winding canal "which wends its way through beautiful scenery representing places of interest in Venice, the beautiful Italian city."[16]

Overholser and Mahan also concentrated on small details that made big differences. They built a new paddock next to the Grandstand, expanded the Grandstand to seat 10,000, and installed a sewer system with modern toilets. They purchased an electrical power plant, wired all buildings for electricity, added 17 arc lamps to light the Grandstand and race track for night events, and installed a telephone system that included fourteen trunk lines and 40 telephones. No project received as much comment in the newspapers, however, as the laying of bermuda grass on the entire grounds;

Poultry competition, at a time when every farm family depended on hens for food and eggs for cash, was a popular feature at early state fairs (Courtesy Daily Oklahoman).

this not only settled dust but also made the "entire grounds appear a well kept park."[17]

The momentum continued into 1910 and 1911 as the list of new buildings lengthened: the Mineral Resources Building, an addition to the Agriculture Building, the Women's and Children's Building, the Cement Exhibit Building, the Bee and Honey Building, the Automobile and Carriage Building, the A&M College Building, and a summer "air dome" theater, constructed for $6,000. The fair board even decided to spend $1,645 to clear 43 acres of brush and timber on the south side of the lake.[18]

Of all the improvements made during this three-year frenzy of construction, the largest and most notable by far was the Livestock and Horse Show Pavilion. In May of 1910 the fair board voted to borrow $7,000 so work could begin on what Mahan called the "finest pavilion in the Southwest." It was to be 170 feet wide by 250 feet long, two stories tall, and constructed of wood, concrete, and steel trusses. It would seat 3,600 people around an arena measuring 80 by 200 feet, with two rows of 56 boxes, each holding six people. In design, it was similar to pavilions at the Texas State Fair, the Indiana State Fair, and the National Feeders and Breeders Show at Fort Worth. By the time it was completed, the cost would exceed $40,000, making it the single largest investment in the fairgrounds to that time.[19]

To Overholser, Mahan, and the board of directors, the Livestock Pavilion and the emphasis on horses and cattle was well worth borrowing against future revenues. The livestock industry, which initially took a back seat to cash crops such as wheat and cotton, had become increasingly important to the economic-well being of Oklahoma City, especially since the announcement of Oklahoma City's first large-scale industry, the packing plants.

In 1908 officials of Nelson Morris and Company, a large Chicago packing house, had offered to build a $3 million processing plant in Oklahoma City if leaders would provide a $300,000 cash bonus. Members of the Chamber of Commerce—led by many of the same men who had supported the State Fair, such as Classen, Colcord, Lee, and Jones—responded overwhelmingly and used subscriptions to pay the bonus and purchase a 575-acre block of land southwest of town that became Stockyards City. The next year another packer, Schwartzchild and Sulzberger, announced that it too would build a new plant if a bonus was provided. Again, the Chamber met the demands. By 1911 both plants were up and running, a $6 million investment that employed more than 4,000 workers.[20]

The economic impact was immediate. The telephone company announced new investments of $2 million, while Oklahoma Gas and Electric pledged to double the output of its generating plant with an investment of $600,000. Within two years the value of goods manufactured in Oklahoma City increased from $7 million to $17 million, retail sales soared from $16 million to $25 million, and

"$20,000 in Stock Prizes" as well as "Reduced Rates on all Railroads" were promises meant to lure rural farmers and ranchers to the fair (Untitled clipping from State Fair Archives).

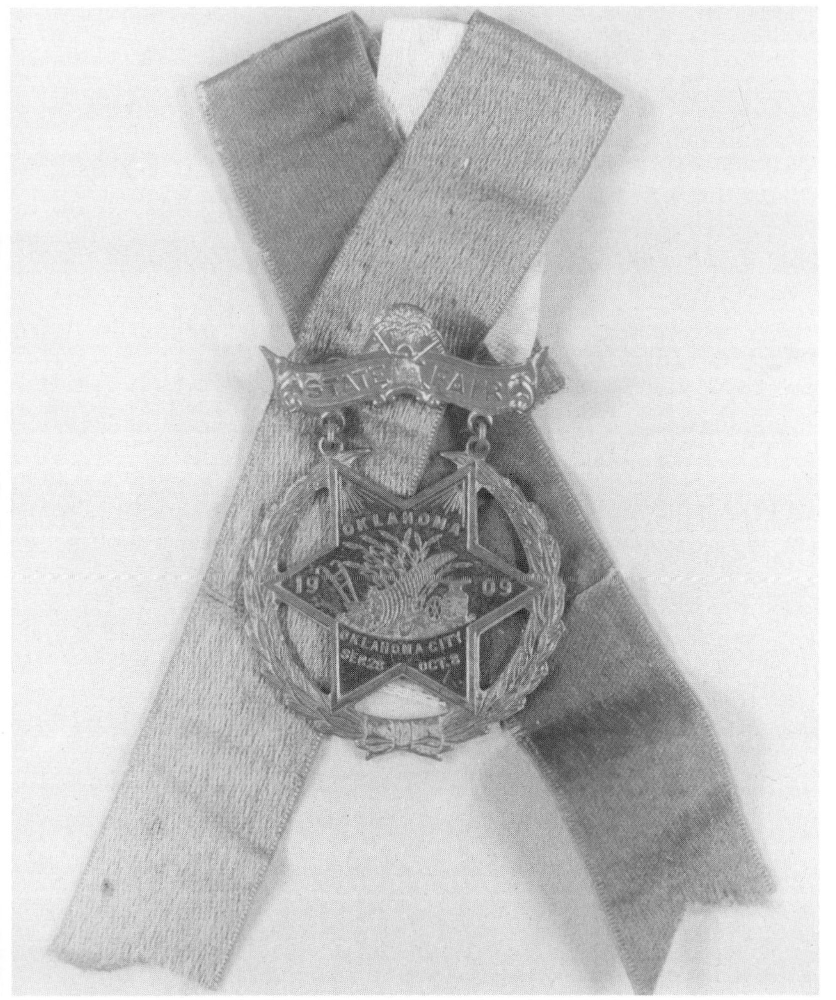

First place ribbon from the 1909 fair (State Fair Archives).

bank clearings leaped from $52 million to $122 million. With all of this economic growth fueled by the packing plants, it was no accident that the fair board voted to showcase the livestock industry with a grand new arena in 1910.[21]

Even before the packing plants were built, farming and ranching had been gaining increased attention from Overholser, Mahan, and the fair board. It was a matter of self-interest. Oklahoma City's economy, then and for years to come, had to rely on the production of farmers and ranchers. The city was too far removed from the oil fields around Tulsa to get direct benefits from exploration, production, or refining. It was cut off from the coal fields around McAlester and the attendant industrial production. And it was far to the west of the new state's most populated counties. Oklahoma City's only advan-

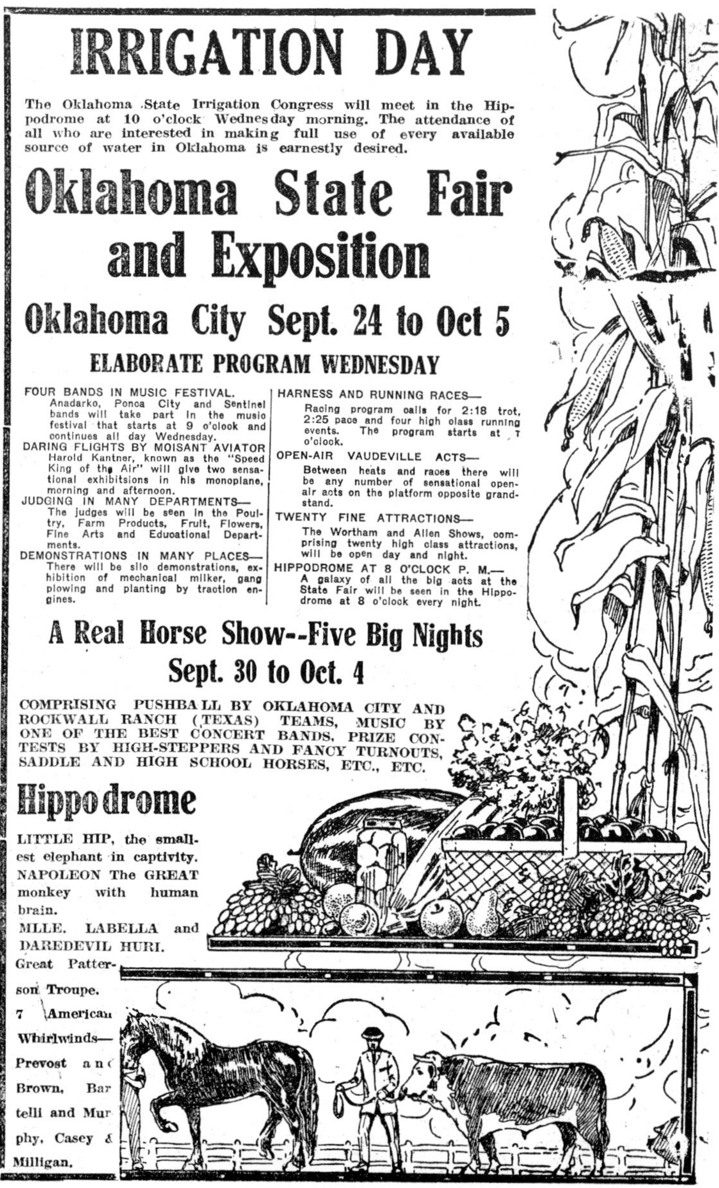

Scientific farming, a major initiative of the fair, was illustrated by the headline of this advertisement that touted "making full use of every available source of water in Oklahoma" (Untitled clipping from State Fair Archives).

tage, other than transportation, was the rich farmlands
that stretched through the middle of the state on a northeast to southwest diagonal. If the city's businessmen
were to prosper, the area's farmers and ranchers had to
produce larger and better crops.

This recognized importance of agricultural production coincided with a new emphasis on "scientific farming." Throughout the western world, it was the age of
scientific and technological exploration, with a common
belief that man could overcome any challenge if only he
applied the powers of logic and experimentation. The
results included immunization against disease, heavier
than air flight, internal combustion engines, wireless
telecommunication, and the discovery of X-ray. It was a
natural extension that these same principals of logic and
experimentation should be applied to agriculture, the

County competition, encouraged at the State Fair of Oklahoma, promoted greater production and better attendance from rural areas. Diversification was the theme of the Blaine County exhibit, as seen in the caricature of the "cow-sow-hen" (Courtesy Daily Oklahoman).

bedrock of the American economy.

The State Fair of Oklahoma played a major role in the
new age of scientific farming. Where else, agricultural
leaders asked, could they reach such a large audience of
farmers and ranchers and preach the virtues of conservation, organization, and efficiency. "It goes without say-
ing," wrote one editor, "that any organized movement or
system of teaching agriculture will keep the rising generation of farmers' children on the farm and at the same
time make better farmers of them than their fathers were
before them....This, in the end, will be a blessing to the
nation at large."[22]

To Overholser, Mahan, and the fair board, the traditional agricultural displays of counties and individual farmers were no longer enough. While such exhibits satisfied the need for boosterism and even helped draw large crowds from rural areas, they had little impact on the need for more education and the application of scientific farming. In 1910 the fair board found an efficient way to supplement those popular attractions—the first State Fair Agricultural School.

Under the supervision of professors from Oklahoma Agricultural and Mechanical College, the school in effect was a 12-day fall camp for boys age 14 to 18. Each county could send two boys, selected by the county superintendent of schools and the Farmers' Institute. Tuition was free, but the boys had to arrange for their own transportation, meals, and bedding. Cots were provided under a tent on the grounds.[23]

According to Professor T.M. Jeffords, the school was for "serious study and observation." Each morning, instructors from the A&M College conducted demonstrations on soil judging, planting techniques, and crop diversification. A model dairy was installed where the boys were instructed in cream separation, testing, and churning. A herd of purebred livestock was shipped in from the college in Stillwater with specimens of Jersey, Short Horn, Angus, and Hereford cattle, Shropshire and Merino sheep, Berkshire, Poland and Duroc swine, and Percheron and standard breed horses.

Each night professors conducted illustrated lectures on fruit tree cultivation, soil conservation, and veterinary science. For variety and exercise, one hour each day was devoted to track and sport events, while most afternoons were free time so the boys could see the fair. Mahan even arranged for free streetcar passes so the boys could visit points of interest in the city, such as the packing plants.[24]

The first State Fair Agricultural School was so successful the fair board built a new structure for the program during the off season. Called the Oklahoma A&M College Building, it was 50 x 100 feet in size, with a concrete floor, electric lights, and a fully equipped cooking room.

Opposite page: "Interim Events" such as summer horse racing utilized the fairgrounds throughout the year (Courtesy State Fair Archives).

This display touted the variety of crops grown in Seminole County, an important aspect of "scientific farming" promoted at the State Fair of Oklahoma (Courtesy Daily Oklahoman*).*

This complimentary pass to the 1910 State Fair of Oklahoma was granted to John Hubatka, a member of the city's first police force in 1889 and later chief of the Oklahoma City Police Department (State Fair Archives).

In 1911 the program was expanded to include one girl from every county. The new curriculum ranged from domestic science and art work to physical training and games. The girls slept inside the building, while the boys remained in the tent dormitory.[25]

The importance of the agricultural school and its emphasis on bigger and better farms was the subject of a speech by Governor Robert L. Williams when he addressed the students at their awards assembly in 1916:

> Don't imagine that all town people are prosperous. There is more actual squalor, ignorance, and misery in the cities than in the country. Why should a boy or girl wish to leave the farm to come to hot city pavements? Why should anyone leading a comfortable, independent existence on the soil want to become a slave in the machine of the city? I tell you, boys and girls, that the modern farmer is the most independent man in the world—and should be the happiest....The man, the boy or girl, who raises corn, pigs, poultry, wheat and other food is engaged in a service which not only enriches himself but which makes possible the conduct of the government and continuation of human life and civilization.[26]

To Williams and the leaders of Oklahoma City, a healthy farm economy depended on motivated and educated young farmers, and a healthy farm economy meant progress and prosperity for everyone.

The increased emphasis on scientific farming at the fair was not limited to youth. Even the county displays, once symbols of "freakish exhibits, such as the tallest stalk of corn or the largest beet and greatest variety of stuff grown on one farm," were reoriented with scientific organization and instruction. This transition peaked in 1915 when the Oklahoma Legislature passed the Free Fairs Bill, providing state support for organized agricultural competition from township and county levels to the State Fair.

By the fall of 1916, 30 counties had organized under the free fair law, and in a few counties there were as many as 20 township contests. Each level of competition was under the supervision of the county agricultural extension agent, who was paid from a combination of county, state, and federal funds. His duty, according to one advocate, was to "comb the state for the best livestock and farm products, and interest and educate the folks from the remotest township contest to the survival of the fittest at the state fair, in what is the best and why it is the best." Competition began at the township level during the first week of September, moved to the county fairs the next week, and progressed to the state fair by the third week of the month. As one editor wrote, "with an organization as thorough as this, I am confident that

Features promoted in this 1914 newspaper ad included the horse show, the "Australian riding marvel, May Worth," an elephant show, horse races, a livestock show, a car show, and Louis Disbrow, the world champion automobile racer (Untitled clipping from State Fair Archives).

practically every farmer in the county will receive some helpful influence for better farming or a better grade of livestock."[27]

Overholser and Mahan also provided for agricultural demonstrations and instruction for the general public. Since the first fair in 1907, displays of farm implements had been popular attractions that introduced to farmers and ranchers the latest in plows, cultivators, windmills, generators, and eventually, tractors. In 1912 the newest addition to "implement row" was the silo, which, according to experts, had "revolutionized the raising of livestock in northern states." Agricultural extension agents assembled a metal silo, filled it with corn, and demonstrated the advantages and methods of storing and feeding. Seven different styles of silos were on display, constructed from wood, metal, and concrete.[28]

The next year, the fair board set aside one day of the fair as "Implement Dealers Day" with the express purpose of bringing together in one place farmers, dealers, and manufacturers. To Mahan, it was important that new techniques of scientific farming be available to "the many Oklahoma men and women who have not had the time or opportunity to leave home and attend an agricultural school." With 40 acres devoted to the displays, there was room to demonstrate gang plows, planting by "traction engines," and the many ways "to increase farm outputs at the minimum of work and...add to the comforts of farm life."[29]

The principles of scientific farming were even applied to horticulture through demonstrations of selective cultivation, pruning, and spraying of apple trees. The previous year, A&M professors had applied such methods to 129 trees in an orchard near Stillwater. When harvested, the sprayed trees produced $400 of fruit; the trees not sprayed produced apples worth less than $25. The display also included technical leaflets offering more information on the scientific approach to the management of orchards.[30]

Still another demonstration project was the model dairy installed in 1916. Using a dozen cows of assorted quality, USDA agents provided two lectures daily to explain the virtues of balanced rations, sanitary handling of milk, and the modern design of milking facilities. Milk was taken to a "model dairy house" where it was cooled and strained. The demonstration ended with different methods of testing for butterfat and the sanitary processing of dairy by-products.[31]

In Oklahoma, where 90 percent of all rural families owned at least one milk cow, such demonstrations were more than just curiosities; they were valuable lessons designed to improve living conditions and increase economic output. This same combination of education and

Automobile shows became a popular attraction at the fair as early as 1908. By 1910 there were more than 70 automobile dealers in Oklahoma City, providing a new source of income to the fair (Courtesy Oklahoma Historical Society).

entertainment was found in the fair's most popular new headline attraction, the Horse Show.

Held in the new Livestock Pavilion for the first time in 1910, the Horse Show would have, said Mahan, "the dignity and importance of a separate and well defined feature...and will consist of about every event which exhibition horses may enter, interspersed with band concerts and vaudeville acts." Every evening during the fair, sold out crowds watched competition between saddle horses, Percherons, heavy horses, light horses, hackney horses, and a host of other classes. There was even a troupe of "educated" Shetland ponies that performed under saddle, in four-, six-, and ten-horse harness, and in chariot racing.

The Horse Show was tempered with a variety of entertainment acts, and as Mahan promised, "there was not an idle moment." Concert bands filled the pavilion with music, while vaudeville acts performed between competitions. At the first Horse Show, the headliner was the Bedini Family, a troupe of three acrobats and one Collie named Ulo, who had a "vaulting equestrian act...making simultaneous leaps from the ground to slender backs of thoroughbred running horses." Other performers included the Four Ishikawas Japanese troupe, "the world's greatest equilibrists," the Zamora Family of trapeze artists, and Cordua and Maud, a "clever team of acrobatic artists known as the Craze of Europe."[32]

For the next two years the Horse Show grew in size and stature. Billed as the highlight of the entire fair, the Horse Show of 1911 included 20 general divisions, 82 different contests, and a total purse of $6,813. There were gaited horses, walk-trot-canter horses, cowboy riders, and classes for women and children. Each added to the aura of what one writer called a "society event." Social columnists from local newspapers covered the shows and reported who was attending with whom, and who was wearing the latest fashions of dresses and hats.[33]

Enthusiasm for the Horse Show was complemented by the popularity of horse racing. From 1907 to 1913, the races were featured attractions, drawing large crowds that convinced the fair board to expand the Grandstand not once, but twice, until it seated 10,000 people. Not even the disastrous races of 1910, when one jockey was killed and two riders were seriously injured, could dampen the appeal of the ponies.[34]

Ironically, as enthusiasm for horse racing grew in popularity, opposition to gambling on the races grew even more quickly. As early as 1909 editorials had appeared in local newspapers denouncing the impact of bookies and pool gambling. "Oklahoma is one of the very few states of the central West," wrote one editor,

By 1917 the popularity of horse racing was declining, the victim of the ban on legal betting. To maintain crowds, Mahan added automobile racing (Untitled clipping from State Fair Archives).

"You owe it to your children," read this advertisement for the fair, because "they deserve a vacation that will be nothing short of a liberal education, to say nothing of the world of fun they'll have" (Untitled clipping from State Fair Archives).

"which permits betting on horse races. There were eighteen or twenty professional bookmakers at the fair last week, all doing a thriving business....Oklahoma has much to lose and absolutely nothing to gain in this respect."[35]

In 1911 a bill was introduced in the Oklahoma Legislature making gambling on horse racing illegal. Although the bill died in committee, Governor Lee Cruce took up the issue and lobbied for a way to prohibit what one writer called "the nefarious race track gambling and its accompanying trail of crime, disaster, embezzlement, distress, and general crookedness."[36]

While Cruce pushed for a law banning all gambling at the tracks, Mahan and the fair board experimented with parimutuel machines, a system imported from France and used at tracks around the world. First rented for the summer races in 1912, the machines were described as "the one and only system now in use which is an absolutely fair one." Instead of bookies setting odds and holding money, the public determined the odds through the pattern of their bets, with five percent of the total pool deducted for purses and the remaining amount paid to the lucky bettors. According to reporters at the races, the public responded enthusiastically to the new system.[37]

Despite the use of parimutuels, the Legislature followed Cruce's direction and passed a bill outlawing betting on horses. When three bookmakers tried to set up at the 1913 state fair, the county attorney prosecuted and convinced the judge to issue an injunction. Two days later, on September 24, 1913, the fair board voted to "ban all forms of betting in and under the grandstand during the state fair." Although horse racing would continue at the fair for many years, the prohibition on gambling limited its popularity. Never again would the races be the major draw of the State Fair of Oklahoma.[38]

Fortunately for Mahan and the fair board, a new, even more exciting speed attraction was capturing the public's attention: automobile racing. Part of the appeal was the novelty. The first automobile races dated only to 1895, and those contests had been simple tests of endurance on the open road, usually between cities. The first grand prix races, held on enclosed courses, followed in 1904. Even the Indianapolis Speedway, the most famous of all circular tracks, was not built until 1909. By 1913 and the prospect of a diminished horse racing card, the fair board was ready to experiment with the new sport.

In August of 1913 Mahan negotiated a contract with the Louis Disbrow Racing Team, led by the recently crowned "world's circular dirt track racing king, Louis Disbrow." The team, consisting of seven cars and five "auto pilots," did not come cheap. For two days of rac-

By 1915 Oklahoma City was the largest city in the state, with a diversified economy that depended on the promotional benefits of the State Fair (Courtesy Oklahoma County Metropolitan Library).

ing, payment was to be 40 percent of gross receipts from the gate, grandstand, parking space, and infield concessions. The investment would prove to be a wise one.[39]

The top three drivers were all record holders. In 1912 Disbrow, driving a 290 horsepower Jay-Eye-See as well as the lighter, faster Simplex Zip, had broken seven track records set earlier by the legendary Barney Oldfield. The most experienced driver was Wild Bill Endicott, driving a car called Tornado, described as "one of the fastest 450 cubic inch cars in America today." The other record holder was the "flying Californian," Joe Nikrent, driving a Case he had piloted in the most recent Indianapolis 500. According to advanced billing, he was in second place when two flat tires led to a crash. On the final two days of the 1913 State Fair of Oklahoma, the "auto daredevils" thrilled large crowds as they raced around the half mile track at speeds up to 60 miles per hour.[40]

Mahan, facing the unknown impact of the ban on gambling at the horse races, added another automobile event billed as the "world's most dangerous sport." It was auto polo, played on a field 350 feet long and 125 feet wide, with two teams of racing cars, each carrying a driver and a mallet man who tried to get the ball through the opponent's goal. "The racing automobiles," wrote one reporter, "must be able to withstand collisions, must not be incapacitated by climbing a few fences, and must be so mechanically arranged that should they turtle they can be righted and put in running order in twenty seconds." A constant speed of 30 to 40 miles per hour was maintained during the competition.[41]

Automobile racing grew progressively more popular each year after the inaugural events in 1913. Mahan, who had been elected secretary of a national auto racing association, responded with bigger and faster races. He opened the field to what he called the "younger generation," noting the death of Bob Burman, the retirement of Louis Disbrow, and the "refusal of Barney Oldfield to again jump into this chance-taking end of the speed game." Entrants in 1916 included the team of Eddie Hearns and Louis LeCocq, who drove Briscoes; Billy Carlson, who piloted a big Maxwell; and "Red" Schaefer, who immediately took his Chevrolet to a local garage to repair damage done in a race in Colorado. The field also included Cliff Woodbury in a powerful Deusenberg, Sig Hughdahl at the wheel of a Mercer, and John Raimey,

A splendid display for the first time of standard 1916 motor cacrs will be one of the features of the State Fair.

Various models in the different makes will attract the interest and attention of all visitors as well as prospective buyers.

Touring cars, runabouts, roadsters, sedans, limousines will be exhibited by the various dealers. Electric cars will also be in evidence.

More than 25 different makes—several models of each—will await your inspection.

The following will be represented:

CHALMERS	ARGO	BRISCOE	GRANT SIX	WILLYS-KNIGHT
HAYNES	INTERSTATE	SAXON	REO	STUTZ
HUDSON	STUDEBAKER	MOON	MAXWELL	PAIGE
BUICK	OVERLAND	OAKLAND	MARION	CHANDLER
HUPMOBILE	KING	DODGE	CADILLAC	DETROIT ELECTRIC
		JONES-SIX	NATIONAL	

In Automobile Bldg.

To the North From Inside Main Entrance Gates

The Moon and Jones-Six cars will be found north of the Automobile building the Interstate and Marion to the south and the Chalmers l'ne in the northwest section of the Exposition building.

AUTO RACES

General Admission 50c.
Grand Stand 25c.

More than 25 different makes of vehicles were featured at the State Fair in 1915 (Untitled clipping from State Fair Archives).

who drove a "speed wagon that is touted as one of the fastest light car racing machines ever revamped from a Ford car." A special entrant was Miss Elfreida Mais, a 19-year-old girl billed as the "world's fastest woman driver."[42]

On race day a record crowd of 36,000 people watched the "sputtering, roaring machines as they whirred around the half-mile track." The one-mile time trial was won by LeCocq in 1 minute, 13 seconds. The five-mile match race for $100 and a gold medal was won by Woodbury in a time of 6 minutes, 25 seconds. Other times were 3:56 in the three-mile dash, 1:19 in the Ford mile exhibition, 6:29 in the five-mile derby, and 6:27 in the five-mile sweepstakes. Miss Mais, driving her "Mais Special," navigated the mile in 1:20.[43]

Louis Disbrow was the featured attraction at the 1914 fair, along with his "Jay-Eye-See," the "world's most powerful racing car" (Clipping from the Daily Oklahoman, *September 27, 1914).*

The exploding popularity of automobile racing was reflected in advertising as early as 1913. In broadsides distributed to newspapers, the biggest, boldest type announced three main events, "AUTO POLO," "AUTO RACES," and the "SOCIETY HORSE SHOW." Horse racing, previously the center attraction in all such advertisements, was not even mentioned. Clearly, a new era of entertainment had begun at the State Fair of Oklahoma.[44]

The ability to react to change had been and would continue to be the measure of the fair's success or failure. The years leading up to World War I were ripe with change. First came the shifting economy, so the fair board committed more and more resources to scientific farming and livestock production. Then came the ban on gambling, so the board turned to automobile racing. There was one change, however, that could not be dealt with so easily, and that was state politics.

The issue was the right to be called a state fair. When Oklahoma City business leaders organized their exposition in 1907, there was no real competition for the title "state fair." Guthrie, with a population stalled at 10,000 people, could not or would not muster the resources to try, while Tulsa, a young oil boom town with lots of promise, did not even have a local free fair until 1913. The prime contender, surprisingly, was Muskogee.

The people of Muskogee justifiably had high aspirations. Culturally, the greater Three Forks area around

Above and opposite page: This panorama photograph of the 1919 fair indicates the diversity of displays and the public fascination with automobiles (State Fair Archives).

Muskogee was the "cradle of civilization" in the former Twin Territories, home to some of the state's brightest writers, artists, poets, and musicians for more than half a century. Politically, the city was a dominant force, the home of both Robert L. Owen, the renowned U.S. Senator from Oklahoma, and Charles Haskell, the first governor of the state. And the traditions of political statecraft ran deep among the people of Muskogee, nurtured through a century of fighting with the federal government over removal, tribal sovereignty, and ultimately, allotment and the very survival of native traditions. It was no accident that the president, vice-president, and secretary of the Oklahoma Constitutional Convention all had come from the old Indian Territory.

In January of 1910, riding the crest of the same economic boom affecting Oklahoma City, the business leaders of Muskogee had announced plans for an annual fair that would be "the biggest in the Southwest." They formed a corporation capitalized at $50,000, rented an 80-acre park near town, and quickly publicized plans to build a one-mile track, a $10,000 grandstand, and several exhibit buildings and barns that would accommodate up to 200 horses. After all, they noted, Muskogee had been home to the first fair in Oklahoma, and it was currently

the fastest growing city in the most densely populated part of the state.[45]

Their ambitions were soon clear enough. On the third day of the 1911 session of the State Legislature, a bill naming the Muskogee exposition the "State Fair of Oklahoma" was passed on the floor of the House. The Oklahoma County delegation, in a deal that guaranteed Muskogee support for removal of the capital from Guthrie to Oklahoma City, did not fight the issue. On the Senate side, however, the bill was momentarily confined to the Agricultural Committee, and it took lobbyists from Muskogee several weeks to move it out and onto the floor for a vote. It was approved by the Senate, but not before Governor Charles Haskell, a native of Muskogee, had left office. It would be up to his successor, Lee Cruce, to sign the bill.[46]

Cruce, a conservative Democrat from Ardmore, vetoed the fair bill. "Personally," he said, "I am opposed to the government undertaking to do a thing that an individual or corporation can do as well. Fairs are largely private enterprises and as such should be managed apart from the government." He also cited the long-term financial commitment that the state would have with a fair. "Several sections of this bill," he noted,

"leave no doubt that it is the intention of the promoters of the proposition to ask future legislatures to give financial aid." Too late for an override, the bill was rewritten during the next legislative session, and once again it passed both the House and the Senate. Cruce, still in office, killed it a second time with another pocket veto.[47]

The cause was resurrected in 1917 when Muskogee boosters, led by editor Tams Bixby, made an all-out effort to pass what was referred to as the Muskogee Fair Bill. This time, no longer needing Muskogee's support for moving the state capital, the leadership in Oklahoma City publicly attacked the bill. The *Daily Oklahoman*, published by E.K. Gaylord, did not mince words. "This bill, if it becomes a law, will make the Muskogee exposition the official state fair of Oklahoma. The state will manage it and finance it and every premium awarded will, in effect, carry the seal of the state. It will put the Oklahoma City state fair out of business."[48]

Opposition also came from towns to the west of Oklahoma City. The editor of the Chickasha newspaper stated flatly that his county "would not take an exhibit to Muskogee owing to the distance and expense." Even the editor of the Bartlesville newspaper questioned the need for an official state fair designation. In his opinion, any community had the option of building a fair. The designation as a state fair, the editor believed, should be based on merit, not politics.[49]

Despite such opposition, the Muskogee Fair Bill passed both the House and the Senate in mid-February of 1917. It then went to Governor Robert L. Williams, a former Supreme Court Justice from Durant who had previously said he would sign the bill if passed. Before he did, however, he wanted an opinion from the Attorney General on the constitutionality of the law.[50]

Meanwhile, only token resistance came from the preoccupied Oklahoma City fair board, which had limped from one crisis to the next for several years. The most pressing problem in 1917, just as it had been in 1907, was financial. The era of physical expansion from 1908 to 1912 had been funded through loans, not revenues, and the meager profits from one year to the next never seemed to cover the interest, much less the principal. By 1917 the cumulative debt exceeded $174,000 with no prospects of raising the money. Bankruptcy was a real possibility.[51]

Compounding these problems was a crisis in leadership. Jones, the original booster, had long since removed himself from the board of directors and never again became involved with the fair. His successor, Henry Overholser, had taken up the cause with all his might, but he had suffered a paralyzing stroke in 1913. Accord-

He Always Has Been and Always Will Be a Rank Outsider at the Oklahoma State Fair

ing to one close associate, Overholser was formulating plans for a bigger and better fair on the very day of his attack. Mahan did a good job running the daily business of the fair, but no one on the board of directors stepped forward with the commitment and sacrifice so desperately needed. In September of 1915, Henry Overholser, the "father of the State Fair of Oklahoma," died.[52]

Although not recorded in the minutes of their meetings, members of the fair board must have questioned the future of the fair as they prepared for the 1916 exposition. They were deep in debt with little hope of increased revenue, and there had not been a major construction project since Overholser's stroke in 1913. As if that was not enough, a fire blazed out of control on the night of August 21, 1916, only a month before the next fair was to open. The flames destroyed the Grandstand, the Carousel, the Concession Hall, and half of the amusement rides. More importantly, the flames seemingly destroyed the will to carry on.[53]

The fair of 1916 opened on time, but with a half-completed Grandstand. In a way, it was symbolic of the larger issue—too much already built to walk away from, yet too much to complete with the limited resources at hand. After a decade of hope and promise, the State Fair of Oklahoma was at a turning point in its young history. The future of the institution, it turned out, was in the hands of the people of Oklahoma City.

HIGHEST AWARD
COUNTY EXHIBIT
OKLAHOMA
STATE FAIR
AND
EXPOSITION

Chapter 4
ANOTHER CHANCE

For ten years, from 1907 to 1916, the State Fair of Oklahoma had grown steadily as a private corporation free from governmental support or control. That independence worked well as long as attendance and amusements generated enough revenues to pay for operational expenses and benevolent leaders were willing to invest time and money in a good cause. By 1916 that formula for success had come to a startling, abrupt end.

The threat from the Muskogee State Fair, in the final analysis, proved to be the least of the problems. The upstart exposition was indeed recognized as the "official" state fair, a political victory that survived a hostile attorney general's opinion and a referendum petition that put the law to a vote of the people. But there were no state appropriations for the festival, and the 1917 and 1918 editions of the "east side" fair proved to be little different than those of earlier years. If anything, the political success of Muskogee helped the Oklahoma City fair by stirring up latent passions of civic pride and competition, sentiments that would prove essential in coming months.[1]

The real threat to survival was an inherent weakness in the organizational structure of Oklahoma City's State Fair of Oklahoma. As an undercapitalized private asso-

Opposite page: The highest award for "County Agricultural Exhibit," as it appeared in 1935 (Courtesy Daily Oklahoman*).*

J.H. Everest was a member of the fair board from 1916 to 1932, and served as president in 1926 (State Fair Archives).

ciation, investments in buildings and improvements had been made with borrowed funds even though the organizers probably knew that the size and educational nature of the fair would never generate enough profit to pay the interest, much less the principal. Even the ability to pay operational expenses was threatened by the loss of race track gambling in 1913 and a devastating fire in 1916. To survive, the fair either had to expand its commercial side and become a purely for-profit institution, in essence a glorified amusement park, or retain its educational and promotional personality and find a new, broader base of financial support.

Other state fairs in the Midwest and Southwest had faced the same dilemma. Some, such as the Kansas and Iowa fairs, secured state support through the appropriations process, but typically, once one city in those states received funds for a fair, other cities demanded and received similar appropriations, diluting the significance of being a state fair. The most successful solution had proven to be direct city support, a formula that had worked wonders at the one institution held up as a model for the Oklahoma City fair since 1907—the State Fair of Texas.

Amazingly, the circumstances leading to the financial crisis in Oklahoma City were identical to the problems that had almost destroyed the fair in Dallas. As in Oklahoma City, a destructive fire had gutted the Exposition Building, the Music Hall, the Poultry Building, and several barns in 1902. Then came a crushing blow to the primary source of revenue—horse racing—when the Texas legislature banned pool betting and bookmaking. After the dismal failure of the 1903 fair, a syndicate of investors from St. Louis had offered $125,000 for the institution and announced that they would subdivide the fairgrounds into a suburban housing addition.[2]

Before selling and thereby forfeiting the title of state fair to another Texas city, the stockholders in Dallas proposed a partnership with the city based on a plan that had worked well in Toronto for more than 20 years. The city, through a special vote, would raise $125,000 to purchase the grounds; $85,000 would pay off debts, and the remainder would be used for capital improvements. The city would own the grounds as a Fair Park, while the fair board would have control of the grounds during the fair each fall. After an effective promotional campaign, the people of Dallas approved both the management and financial plans by a vote of 2,531 to 415.[3]

On November 24, 1916, the State Fair of Oklahoma board of directors asked Mahan to investigate the "Dallas Plan." Any reluctance on the board was washed away on January 10, 1917, when another fire destroyed the Expo-

In the Women's Building county
exhibits lined the wall on the left
while canned goods were judged at
the tables (Courtesy Daily Oklahoman).

sition Building, the sentimental heart of the original
fairgrounds built in 1907. Faced with mounting debts, a
partially completed Grandstand, and the loss of major
buildings with no visible means of replacement, the
stockholders voted on February 26, 1917, to offer the
fairgrounds and all improvements to the City of Oklahoma City for $300,000.[4]

The financial arrangement, designed to attract public
support, was straightforward. The stockholders, by their
own vote, would receive nothing, in effect making their
previous investments one large donation. The quarter
section of state school land, which some appraised as a
$100,000 value, would be purchased for $34,000. A small
bank loan would be paid, and the $172,000 debt to the
estate of Henry Overholser would be settled for
$100,000.[5]

The rest of the money would go for improvements.
The partially completed concrete Grandstand would be
finished at an estimated cost of $38,800, including a steel
roof. A new Liberal Arts Building, designed for commercial exhibits as well as a home for the culinary, fine arts,
textile and educational departments, would be constructed for $100,000. Other construction projects in-
cluded a Women's Rest Cottage for $5,000, a $5,000
remodeling of the Livestock Pavilion to make it suitable
for auditorium and convention purposes, and various
improvements to sidewalks, roads, and fences.[6]

Motorcycle racers tear up the track in
1934 (Courtesy Daily Oklahoman).

The Liberal Arts Building, constructed in 1918 and 1919, had 37,000 square feet of exhibit space (Courtesy Daily Oklahoman*).*

The operating agreement was a bit more complex. The city would own the grounds and buildings, and for eleven months of the year operate the entire site as a city park. Any revenues from events in the buildings during those months would go to the city to pay for upkeep, insurance, and other expenses. Each fall, the city would lease the grounds and buildings to the fair association for 30 days, during which time the fair board would control access to and management of the park. After expenses were deducted, any profits from the fair would be split, 50 percent up to $25,000 retained by the fair board for operating funds and 50 percent allocated to capital improvements on the grounds. No project would be undertaken without full consent of both the city and fair board.[7]

It came down to a question of whether or not the voters of Oklahoma City would approve a bond issue, so the fair board agreed to underwrite a promotional campaign to sell the "Dallas Plan." Full-page advertisements ran in newspapers with headlines such as "What the Proposed State Fair Bond Issue Does for Oklahoma City" and "Is it Worth the Price?" In bold print, promoters simplified the issue, "It figures 27 cents a year—your part of the interest on each $1,000 worth of property." At other times they used scare tactics, "How would you like to go to Muskogee to the State Fair in 1917?"[8]

The entire contract was printed repeatedly in the newspapers with simplified versions running more often.

Letters of endorsement from ministers, businessmen, and politicians were solicited and publicized. L.D. Kight, director of the Real Estate Exchange, wrote "I myself was against voting any more bonds of this nature until I read the astounding offer and realized its immense possibilities for the city." M.D. Scott, owner of a large furniture store, added, "There is no other institution that is of such importance to the retail interest of Oklahoma City as the Fair." Public endorsements were received from the Retailers Association, the Real Estate Board, the Ad Club, the Salesmanship Club, the Rotary Club, the Chamber of Commerce, and "every minister in the city."[9]

Editorials echoed those sentiments. The *Oklahoma City Times* viewed the deal as an "Investment in Parks," an opportunity of "obtaining another park at a bargain price." The *Oklahoma City News* saw it as a challenge: "If this community rejects the bonds next week, we'll have no more State Fair at Oklahoma City....The sentiment of the spirit of this community is being tested." The *Daily Oklahoman* pointed to the general importance of the fair: "It has won recognition as one of the first class agricultural and mechanical exhibits of the country. It brings thousands of people here. It furnishes wholesome amusement. It provides valuable instruction for everybody. It is one of the great educational institutions of the state and of this city. It is therefore of substantial value to every citizen of this city, not only the property owner, but every man or woman who makes his living here."[10]

The strategy worked. On May 8, 1917, the voters went to the polls and approved the bond issue by a vote of 2,444 to 1,223, almost 2 to 1. Another chance for the State Fair of Oklahoma had begun.[11]

Physically, the fairgrounds took on a new look over the next two decades, first with bond money, then with a combination of city and private funds. To design the major additions, the fair board and city commissioners selected the architectural firm of Layton and Smith, directed by Solomon Layton, one of Oklahoma's most important architects for more than 30 years. He and his partners had already designed landmarks such as the Skirvin Hotel, the Oklahoma Publishing Company Building, Central High School, and the State Capitol. They later would design architectural gems such as the Oklahoma Historical Society Building, the Telephone Building, the Oklahoma County Courthouse, and scores of schools and public buildings in all parts of the state. Layton would leave his firm's imprint on the fairgrounds as well.[12]

First to be completed was the Women's Rest Cottage, a modern-looking structure constructed for $10,000, of

With war raging across the Atlantic in 1917, supporters of the fair park bond issue used a patriotic approach in this editorial cartoon (Untitled clipping from State Fair Archives).

which $5,000 came from the bond issue, the other $5,000 appropriated by the fair board. It was completely fire-proof, made from brick and stucco using a design that hinted at the new style called Art Deco. It had 2,700 square feet of space, with 20 lavatories at one end and a large reception area furnished with easy chairs and beds at the other. A "competent matron" supervised the cottage. Within a few years the exterior would be covered with ivy, giving it a country look.[13]

Next to be completed was the Grandstand, which had been partially built at the time of the bond issue election. It was expanded to seat 7,500 under a new steel roof, with seating for another 2,500 on bleachers to either side. Exhibit booths were finished on the lower floors. Then came renovation of an old horse barn for the Boys Dormitory, which measured 306 feet long by 80 feet wide, room enough to sleep 288 boys and feed 360 boys and girls. It included a lobby and classroom space for the agricultural school.[14]

The Pullman Porters' Quartet, sponsored by the Pullman Company, offered songs such as "Carry Me Back to Old Virginy," then invited listeners to inspect two modern Pullman coaches at the 1931 fair (Courtesy Oklahoma Historical Society).

Meanwhile, the staff completed several small projects. The old office building, constructed in 1907, was moved to provide more space inside the entry gates. The Indian Building, the site of agricultural exhibits from Indian schools since 1912, was moved to a lot between the Mineral Building and the Machinery Building. The fair board also added miles of new concrete walks, painted all existing structures, laid out a parking lot inside the grounds for 3,000 cars, and planted thousands of trees and bushes in a general beautification effort.[15]

The most ambitious project was the Liberal Arts Building, budgeted at a cost of $96,500. As designed by Layton and Smith, it was a one-story brick building, looking much like a school, with 37,000 square feet of space for 110 display booths. There were four large entrances, with the main door on the south side. Even before it was finished in 1919, every booth had been rented by merchants, manufacturers, or jobbers.[16]

With the bond money spent, the fair board turned to outside sources for additional projects. In 1923 the

Opposite page: By 1928 the fairgrounds had a number of permanent amusement rides, a new brick Entrance Building (upper left), and rows of games, food vendors, and sideshows under canvas (Courtesy Daily Oklahoman).

A variety of food vendors set up booths at the 1939 fair, including this barbeque joint with a fixed location at 3405 S. Robinson in Capitol Hill (Courtesy Daily Oklahoman).

board approved a request from the '89ers Association to build a club house on the grounds for their members, who raised more than $10,000 for a one-story brick building with a wide veranda, club room, and restrooms. Completed in the fall of 1925, it became the annual rendezvous for reunions of "Payne's Boomers," the pioneers who had made the land run of 1889.[17]

A similar arrangement was made with the Oklahoma City Saddle Horse Club in 1924. In this instance, the club loaned $6,500 to the fair board to build the stables, with the provision that the facility would be leased to the club for $960 a year. The fair board approved the deal, stating an intent to "stimulate the horse industry and to try to build up one of the features of our fair which for a number of years has been continuously going down hill."[18]

John E. O'Neil served as president of the fair board in 1923 and again from 1926 to 1928, good years for the fair when attendance and revenues were growing (State Fair Archives).

Other than barns, walkways, and streets, the most important construction project between 1925 and 1937 was the 4-H Building, the only major improvement funded by the city after the original bond issue. The need was clear. After 1929, when the Agriculture Building burned, students attending the week-long agricultural school were housed in a converted horse barn that leaked when it rained. Despite the poor facility, the number of students increased every year as the 4-H movement gained momentum. By 1932 there were 50,000 members of 4-H in Oklahoma, with 1,000 attending the fair each year.[19]

Oil money allowed the city to appropriate $50,000 for the building during the worst part of the Great Depression. The Oklahoma City Oil Field, the largest in the spectacular Mid-Continent sands, was discovered in 1928 and had increased production every year for more than a decade. Much of the oil was found under city land, producing enough royalties that the city virtually eliminated taxes and still funded capital projects.

The 4-H Building, completed in time for the 1932 fair, was two stories tall and as long as a football field. The first floor included exhibition space, a kitchen, a dining hall, and a large assembly area that could seat 1,000. The second floor was dedicated to dormitory space with steel bunks for sleeping. At the dedication ceremony, Dr. C.W. Warburton, director of the Extension Department of the United States Department of

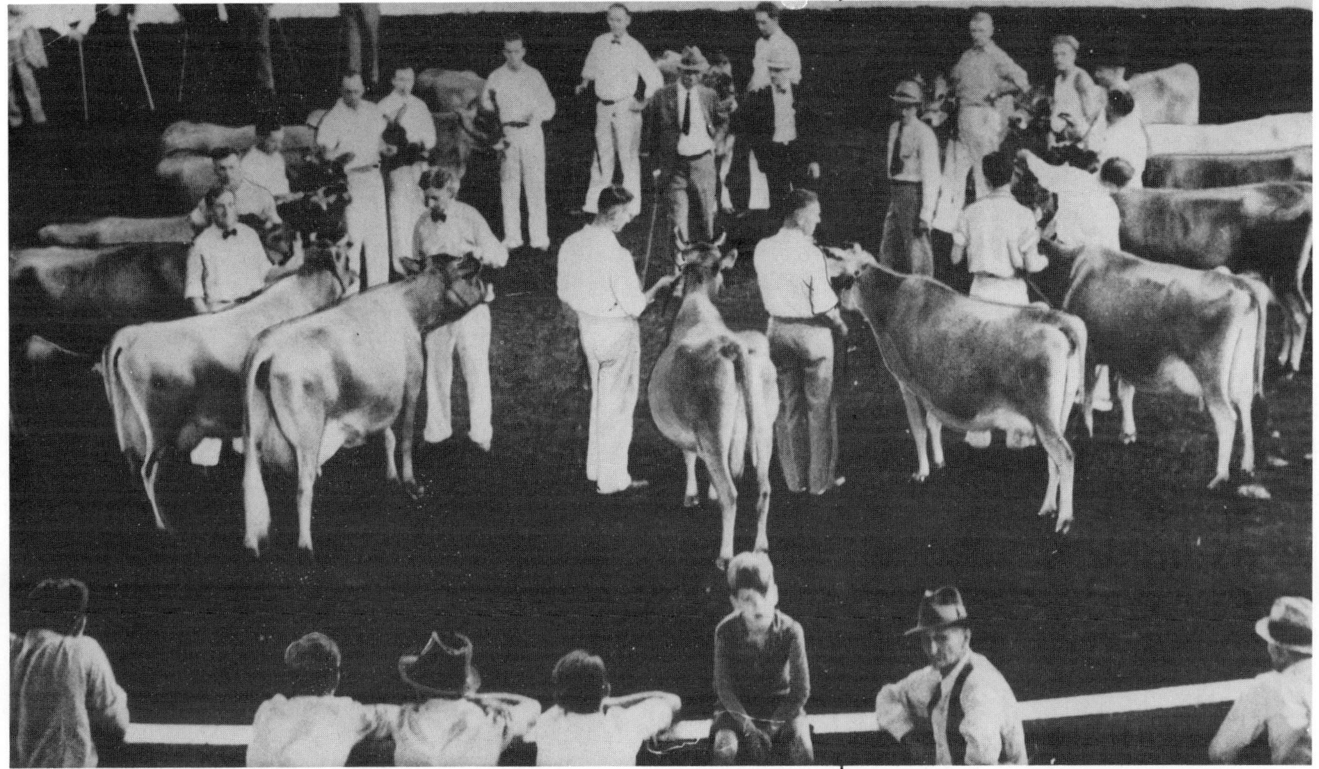

Four-year-old Jerseys, brought to the State Fair from four states, represented the largest class of cattle at the 1935 fair. F.W. Atkeson of Manhattan, Kansas, was the judge (Courtesy Daily Oklahoman).

Agriculture, recognized the significance of the project: "This building," he said, "has no equal any place in the country...none that combine living quarters, exhibit space, school rooms, and other special features under one roof."[20]

Construction of the 4-H Building was proof that the State Fair of Oklahoma remained true to its original emphasis on agriculture. Farm and stock exhibits were still popular in the 1920s and 1930s and actually grew in numbers and quality despite farm consolidation and the growing urbanization of the state. In 1936, one of the worst years of the Great Depression, the agricultural display included 24 county exhibits, 12 instructional exhibits by Oklahoma A&M, 120 farm exhibits by members of Future Farmers of America, and 599 animals in the livestock competition.[21]

The farm implement show also grew until it covered 80 acres with new and bigger displays of modern machinery such as tractors, threshers, combines, and the newest invention, mechanical cotton pickers. In 1925 the implement show theme was "adaptability," and exhibitors clearly demonstrated the functional uses of their machinery. Ford proudly displayed 60 attachments for its tractors and listed 236 "approved" uses to which their machinery could be safely and efficiently used. The trend was towards smaller and more efficient size. One of the most popular machines at the 1925 fair was a Gleaner combine, which "rides straddle a Ford and cuts a swath eight feet three inches wide." Threshers, too, were smaller, with 20-inch cylinders attached to power take-offs on the tractor.[22]

Fair organizers used the agricultural and technological theme to promote the emerging motor truck industry. Before World War I, there had been no commercial trucking industry. Railroads, serving rural communities fed by wagon roads, still carried virtually all overland freight, even on short hauls within the state that were not

A.Y. Owen, who later gained fame as a photographer for Life Magazine, *climbed an oil derrick south of the fairgrounds to take this night image of the fairgrounds in 1936. A fireworks display erupts in the Grandstand infield (right) while rides on the midway (left) are packed with people (Courtesy* Daily Oklahoman).

profitable. Light, fragile, and underpowered, motor vehicles of that era could not compete.

The war altered that reliance on rail forever. Mobilization and the appropriation of rail capacity for moving men and war materiel caused a transportation problem that was solved only by using bigger and more powerful motor trucks, especially on short hauls between cities. When the war ended, the companies that had supplied the trucks and tires, as well as the men who had found a lucrative new business, decided to continue and expand their ventures. Harvey Firestone, sensing a new market for his tires, inaugurated the "Ship by Truck" movement to promote the infant motor freight business.[23]

The fair board designated Saturday, September 30, 1919, as "Ship by Truck" Day at the State Fair of Oklahoma. The festivities started with a parade of more than 250 trucks from the downtown business district to the fairgrounds, where they assembled in the demonstration field. Several truck makers were there, such as the Federal Truck Company, which advertised that pioneer merchant A.M. DeBolt had recently purchased two 3-ton Federal trucks, one with dump body and hydraulic hoist and another with long wheelbase and roller body for handling lumber. The Maxwell Car Company offered a two-ton truck for $1,085, promising it would pay for itself in less that a year. Ford Motor Company featured a light one-ton truck that sold for $590, "just the thing for Oklahoma's rocks and ruts, dust and dryness, heat and heavy roads."[24]

In the best tradition of technological innovation and economic development, fair supporters touted the truck exhibit as an important step forward for city folk as well as farmers and ranchers. "The motor freight hauling business," wrote one editor, "brings together the farmer and the city consumer." Flexible daily delivery of produce to town via truck meant greater efficiency and less waste. Quicker distribution of manufactured goods to rural customers on regularly scheduled truck routes meant greater sales and cheaper prices. By promoting

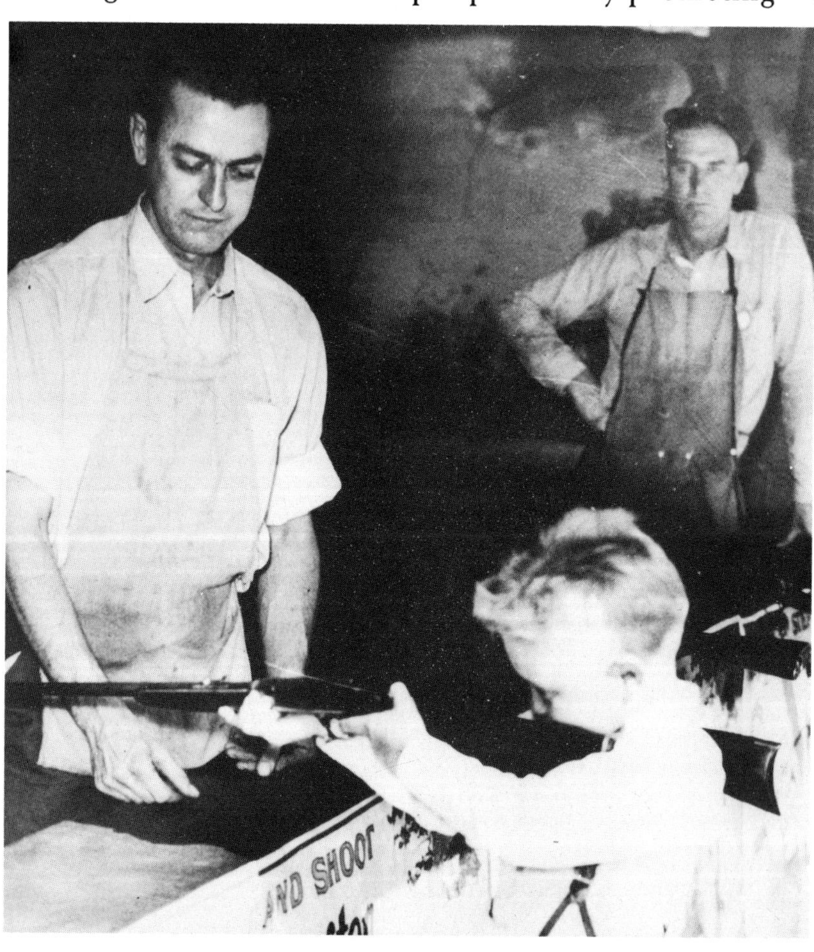

One of the earliest games to hit the midway was the shooting gallery, seen here in 1935 (Courtesy Daily Oklahoman*).*

the truck industry, the State Fair of Oklahoma was contributing to the economic growth of the state.[25]

As trucks became more sophisticated and the motor freight industry matured, truck displays were incorporated into the annual Auto Show, which became more popular each year as "car fever" spread across the country. The earliest display of automobiles at the fair dated to 1909 when Oscar G. Lee, one of the founding fair board members, built a small exhibit for what he called "the largest and most expensive display of automobiles ever placed in a state fairgrounds in the West." His firm, the Lee Motor Company, featured 12 "high class machines," including 5- and 7-passenger touring cars, runabouts, taxicabs, and commercial machines manufactured by Pullman, Knox, and Studebaker.[26]

Encouraged by that showing, the fair board constructed an Automobile and Carriage Building in 1911. The first year, the 20,000 square foot building was divided roughly half for horse-drawn carriages and half for automobiles; within four years automobile dealers occupied the entire building plus a row of tents along the walk toward the main gate. Carriages would not be shown again at the fairgrounds for many years, and then only as antiques.[27]

In 1916, only one year after the first Model Ts were assembled, the Auto Show became a major event for the first time. Wednesday was declared "Automobile Day," with young ladies lined up just inside the exhibit building where they pinned small felt pennants on lapels and coats. As one reporter said, "the pennants shout to the world the merits of various brands of cars." And what a selection there was: Cadillac, Chalmers, Mitchell, Maxwell, Studebaker, Overland, Hudson, Dodge, Paige, Saxon, Hupmobile, National, Chandler, Oakland, Detroiter, Grant, Briscoe, Dot, Reo, Chevrolet, Apperson,

Oklahoma City, looking east from Walker along N.W. 2nd Street, was booming in 1928 with new buildings and a robust economy. Seen in this panorama are the Telephone Building (on left with flag on top), the Petroleum Building (tallest in the center), and an expanded Skirvin Hotel (at the far end of the tracks) (Courtesy Oklahoma Historical Society).

The Petroleum Building (left), Ramsey Tower (center), and First National Bank Building (right), were monuments to the boom and bust cycle in Oklahoma City. In 1930 the town seemed to be on a winner's roll; by 1931 the effects of the Great Depression proved otherwise (Courtesy Oklahoma Historical Society).

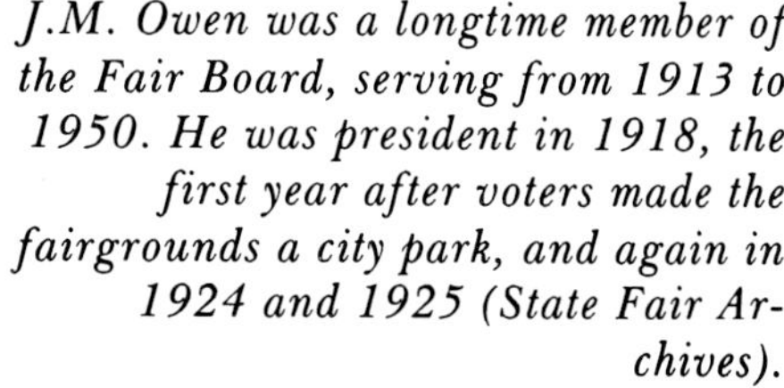

J.M. Owen was a longtime member of
the Fair Board, serving from 1913 to
1950. He was president in 1918, the
first year after voters made the
fairgrounds a city park, and again in
1924 and 1925 (State Fair Archives).

Jackson-Allen, Oldsmobile, and the most recent make, Ford. On Wednesday evening, the Automobile Dealers Association hired an orchestra to entertain the crowds in style as they leisurely inspected the merits of each manufacturer.[28]

As one of the most popular features at the fair, the Auto Show would continue for many years with only slight variations. Occasionally, a new wrinkle would appear, such as in 1920 when Mahan added a "Motor Style Show" with women drivers competing for prizes. Young girls, driving cars provided by local dealers, paraded in front of the Grandstand with points awarded for dress, grace, and assurance as well as style of the car. There were nine divisions, the first five for open touring cars and roadsters, the next three for limousines and sedans. The last category was for electrics.[29]

Technology, whether applied to farm production, short haul trucking, or just getting from here to there, was a dominant theme at the fair throughout the early decades of the century. It was the "machine age," a time when Americans elevated inventors and tinkerers to the status of heroes. Men such as Thomas Alva Edison, Orville and Wilbur Wright, and Henry Ford were admired not because they made fortunes, but because they conquered problems and challenges through technological innovation. And of all the conquests, all the giant leaps forward, nothing captured the public's imagination like aviation. The first flight at the State Fair of Oklahoma was staged in 1908, only five short years after the Wright brothers made history at Kitty Hawk, North Carolina. The fair's first ascent into the skies, however,

was not in an airplane; it was under a hot air balloon
piloted by "aeronaut" Professor L.L. Hill and his young
assistant, 12-year-old Frank Rea.[30]

The local press covered the first flight like a major
event, reporting every step. Professor Hill, they noted,
was fitted into a parachute harness attached by a rope to
the bottom of the balloon while hot air from a trench fire
filled the envelope. At the dramatic moment when the
balloon floated up, the parachute rope at first became
taut, lifted its human cargo a few feet, then broke, dropping the aeronaut to the ground. As one reporter hap-
pily discovered, "he was skinned but not broken."[31]

Young Rea was more fortunate on his first ascent.
According to an eyewitness, the takeoff was smooth, and
the balloon went up at an amazing speed until it entered
a cloud. Just as he was out of sight, Rea cut the rope and
"floated earthward" in his parachute. The balloon, said
the reporter, remained in the air for 15 minutes, "longer
than usual."[32]

*Amelia Earhart was the center
attraction at this air show staged at
the fairgrounds in 1933. Her airplane is in the infield (Courtesy
Daily Oklahoman).*

In 1910 the fair featured another flight demonstration, this time the "Goodale Dirigible" flown by Charles J. Strobel of Toledo, Ohio. The craft, 75 feet long and shaped like a cigar, consisted of a hydrogen-filled silk balloon with an undercarriage that looked like an erector set. On one end of the platform a small gasoline engine pushed a propeller, while Strobel moved backward on the works to make the craft rise, forward to bring it back to earth. Direction was controlled by a rudder. Each day Strobel conducted lectures on the dirigible and made two flights, always drawing large crowds.[33]

The first airplane to fly at the State Fair was booked in 1912. Appropriately, it was flown by a home town boy, Frank Champion, who had left his home in Oklahoma City at age 16 to join the Navy. After discharge, he took flying lessons in France and joined the Moisant International Aviators Company. By the time of his first flight in Oklahoma City, he was good enough to impress one reporter who graphically captured the awe and wonder of the new technology:

The Whip was one of the new generation rides at the fair in 1934 (Courtesy Daily Oklahoman).

The big machine ran along the ground for about one hundred feet, then suddenly lifted like a great dragon fly and began to soar heavenward. At a distance of about a mile from the fair grounds, at about 1,000 feet from the earth, the aviator suddenly turned to the right and began a beautiful descent toward the infield of the fair grounds. At a speed of over one hundred miles an hour, with the great motors and compellers turning at a terrific speed and like a great bird, the descent was made until directly in front of the grandstand. Then with a great swoop the aviator straightened his machine, flew past the thousands in the grandstand and when about the level of the roof of that structure, waved his hands at the multitudes and with lightning like rapidity ascended again to the skies until he became a mere speck along the horizon.

After remaining in the air twenty-five minutes he returned again and at an altitude of 3,000 feet, the motor was entirely shut off and the machine, by the aid of the warping wings, elevators, and controls, brought to the field in immense circles, whirls, spirals, and dips that upon lighting upon the field caused the vast audience to cheer and applaud.

In a separate editorial, the publisher described the flight as "the most beautiful and attractive show at the fair. Pumpkin shows and horse races are old," he claimed, "but the flying machine of the heavier than air type is the most marvelous achievement of our age of wonders." Visitors to the fair agreed, he added, "lingering in vast throngs until the last hour to see this beautiful sight."[34]

Over the next few years, the fair board regularly scheduled either flying demonstrations or balloon acts, including the bomber fleet from Fort Sill in 1917. As announced in full-page headlines, the Army had agreed to send eight Curtis R-4 biplanes for a demonstration at the fair. At 9:45 on Saturday morning, the last day of the fair, a large crowd waited expectantly in the grandstands until they saw specks on the horizon. "They quickly grew in size," wrote one reporter, "until they seemed a flock of hawks. A moment later a muffled, distant roar could be heard, and inside an hour after the first sighting, all were sailing and soaring above the fair ground crowds."

The lead plane, flown by Major C.W. Howard, caused the most excitement. He stalled high above the crowd, then tipped forward and dove to within 100 feet of the ground, ending with several loops "as easily as though the machine was held in the air by wires." After demonstrating tactical formations and simulating bomb runs, the pilots landed on a field near the recently completed State Capitol, where they were picked up and paraded into town like conquering heroes.[35]

As the novelty wore off, flights at the fair became

Uncle Sam, representing the good of the nation, was included in a long series of promotional advertisements for the fair in 1917 and 1918 (Untitled clipping from State Fair Archives).

BATTLE OF CHATEAU---THIERRY

A spectacular reproduction in flaming fireworks of the world's most famous battle will be given six nights in front of the grand stand at the Oklahoma State Fair and Exposition from September 22 to 27.

The Battle of Chateau-Thierry—the turning point of the great world war and where American divisions, marines and soldiers first distinguished themselves. The most wonderful scenic fireworks spectacle ever shown.

You mothers and fathers of Oklahoma's fighting sons, who waited and watched at home—this is a performance for you.

Besides this, there is the most elaborate entertainment program ever presented awaiting you for the whole eight days of the fair. Automobile races, auto polo games, daring aviation stunts, harness and running races, music, vaudeville, midway carnival attractions, figure eight, old mill, government exhibits, war trophies, live stock show, agricultural, county, Indian and educational exhibits. Reduced railroad rates.

Oklahoma State Fair and Exposition
SEPTEMBER 20 TO 27

more daring, sometimes with fatal effect. In 1921, before 5,000 spectators in the grandstand, Lieutenant Arthur Emerson demonstrated the dangerous act of wing walking, but with the additional trick of climbing a rope ladder from one plane to another. On the day of the first performance, both planes took off, then approached the fairgrounds in formation with Emerson in the lower plane.

The daredevil climbed out of his seat onto the top wing where he waited for the other pilot to drop the ladder to him. The first three attempts failed, the rope flipping away from the aviator each time. Then, on the fourth pass, the ladder came within a few feet and the frustrated Emerson leaped for it. A reporter captured the terror of the moment: "For a bare second his right hand clasped the rope, then two feet of space appeared between his hand and rope. The agonized moan from the grandstand came. He missed it! An awful hush hovered over the helpless thousands who could not lift a finger to stop his descent. Slowly the air parted and gracefully his body wended its earthbound course....Death was instantaneous."[36]

Eight years later, again seeking to stage a new stunt, the fair board featured an aerial battle in front of the grandstand. The drama began with three companies of infantry marching across the race track infield with an observation balloon floating above. Behind was a battery of artillery. Out of the horizon came an enemy airplane, determined to attack the balloon. The artillery opened fire, followed by a "friendly" airplane that flew out to meet the enemy in a dogfight. After a series of loops and dives, the friendly was shot down, opening the way to the balloon. The battle ended with the balloon on fire and the troops sending the airplane away trailing smoke.[37]

The overwhelming response to the bomber squadron in 1917 and the mock air battle in 1929 was as much a result of patriotic fervor as it was a fascination with flying machines. In 1917 America had entered World War I with the goal of "making the world free for democracy." It was a hugely popular war, and the people of Oklahoma were both curious and enthusiastic about learning more. The fair board, anxious to encourage and satisfy this escalating sense of patriotism, made the war the official theme at both the 1917 and 1918 fairs.

In 1917 one entire exhibit building was filled with the "materials of war and the methods of modern warfare." There were torpedoes, warship and submarine models, airplanes, armored automobiles, and guns of all kind, from Springfield rifles to huge cannon used at the front. Ten acres of ground were transformed into a gigantic battlefield, where troops of the regular army held sham

Ralph Hemphill, the consummate showman, was secretary of the fair for more than 20 years (Courtesy Oklahoma Historical Society).

During the 1928 State Fair of Oklahoma, a historic dramatization set in India reached its climax with a fireworks extravaganza (State Fair Archives).

battles each day in trenches built to look like those in France and Belgium.[38]

In the fall of 1918, as American casualties mounted, the theme of military victory became even more pervasive at the fair. In the textile department, categories were added for knitted army sweaters, socks, and scarfs. The Red Cross offered prizes for the best exhibits of refugee garments made by schools or clubs. In the household articles department, all "frills and furbelows" were eliminated, with prizes offered only for the plainest and most indispensable articles such as tablecloths, napkins, lunch cloths, hot plate mats, buffet scarfs, table runners, sheets, pillow cases, and quilts.[39]

Even the culinary awards adopted the theme of wartime conservation. "No cake should be frosted," read the instructions in the premium catalogue. "Conservation of foodstuffs will be preached—and practiced—at the state fair." Only four kinds of loaf cake were judged, including a light cake made with no butter, no sugar, and at least one-fourth or more of corn or potato starch as a flour substitute. Special prizes were offered for "conservation confection," including potato cakes and cakes made with honey. Such efforts, repeated at a thousand fairs across the country that fall, contributed to the hard-won victory celebrated on November 11, 1918, Armistice Day.[40]

The war even dominated the entertainment schedule. The big attraction at the 1917 fair was "The War of Nations," a pyrotechnical extravaganza that reenacted the capture of a French town. On a stage 450 feet wide in front of the Grandstand, 300 actors played out the human drama of battle in a scale model of Rheims. Airplanes "with living operators" swarmed over the set with machine guns blazing. Artillery "flashed and thundered," followed by troops charging the trenches and fighting in hand-to-hand combat. Armored automobiles and tanks raced from side to side with shells bursting overhead. Finally, the town was set on fire and the cathedral crumbled, leaving the city "one waste of smoking and blackened ruins." The spectacle ended with a $1,500 fireworks show, a patriotic salute to victory, peace, and freedom for all nations.[41]

Since 1908, when Mahan booked the open stage production "Eruption of Vesuvius," historical spectaculars had been popular at the fair. That popularity would grow even more in the postwar era. Mock battles depicting scenes from the Western Front were repeated in 1918 and 1919, with the horrors of trench warfare reenacted with barbed wire, machine guns, massed bombardments, and aerial duels. As enthusiasm for the war waned, other historical adaptations took their place. First came "Montezuma," depicting "one of the world's most historic and terrific conflicts, the storming of the sacred city of the Aztecs and the overthrow of Montezuma by the adventurous Hernán Cortez and his little band of 500 followers." The production was two hours long, with bombs, rockets, colored lights, hand-to-hand combat, and 30,000 square feet of scenery on a stage more than 500 feet long.[42]

A string of historical productions in the 1920s provided a visual tour of the Far East: "Mystic China," set in the ancient Chinese city of Yang Chow Fu; "India," depicting life in Delhi filled with funeral processions, elephants, and royal processions; and "Tokyo," recreating the devastating earthquake and fire that destroyed the imperial city in 1923. "Rome Under Nero," dramatized the "bloody reign of the last of the Caesars," while "1776" provided a patriotic salute to the founding fathers with scenes of the Boston Tea Party, the Signing of the Declaration of Independence, and the Battle of Yorktown.

The last of the big stage productions at the state fair was "The Fall of Troy" in 1927, complete with siege, Trojan Horse, and the sack of the city, all set against a backdrop of fireworks. It was a fitting finale. The golden era of "spectaculars" was rapidly coming to an end, victim to the movie industry that was turning out a

An armored Uncle Sam, carrying a sword emblazoned with "more food," mounts his war horse, the "Oklahoma State Fair" (Untitled clipping from State Fair Archives).

99

This editorial cartoon, extolling the virtues of conservation and production, was sent to all newspapers in the state "free of cost to be used at your discretion" (State Fair Archives).

steady string of celluloid epics such as "Ben Hur," "Cimarron," and the "Ten Commandments." By the 1930s the public's appetite for historical drama was being satisfied at the movie theater, not the fair.[43]

Whereas modern technology hurt the appeal of dramatic stage productions, it provided a major new opportunity for the fair. "Wireless radio" was first demonstrated at the State Fair of Oklahoma in 1912, but that was a noncommercial, technical display to show the "promise of tomorrow." The first commercial radios were exhibited at the fair in 1921, only one year after WKY first went on the air, broadcast live from the garage of electronic pioneer, Earl Hull.[44]

Overnight the exhibit and sale of radio sets became big business at the fair. In 1925 the fair board responded to the trend by booking live broadcasts of "the voice of Oklahoma," station KFRU of Bristow. Throughout the eight day fair, the "Bristow ether family" presented three programs a day from a temporary studio assembled on the grounds. Crowds gathered to see performers such as Jimmy Wilson and his Catfish Band, the Etherial Quartet, and Marshall Van Pool and his Footwarmer Orchestra.[45]

To handle the crowds and exhibitors, the fair board converted the old Minerals Building to the Radio Building. Called "the most striking example of the march of progress to be found at the state fair," the radio exhibit included a crystal studio that "furnishes the fair visitors a glimpse into the mysteries of broadcasting." It also provided space for 31 booths where dealers displayed a wide variety of sets that promised "to bring in more distant stations, tune sharper, and give clear tonal quality and volume to the sounds produced."[46]

Each year the fair offered something new in what became known as the "Radio Show." In 1929 RCA set up a large display that included a reproduction of the original wireless set Marconi first perfected in 1901. The company also displayed a "radio photography set" that captured an image and transmitted it over the air waves to a receiving set that "drew" a picture using dots and dashes. Three years later a true "television" set was included in the exhibit, a mechanical prototype that used a spinning disk to break images into more than 100 dots of light that were projected across the room by electronic transmission. Although the mechanical television later would lose the technological race to an electrical model, visitors to the State Fair of Oklahoma in 1932 must have marvelled at the "march of progress."[47]

Members of the fair board, responsible for the financial and physical success of the fair, were more concerned about the "march of time." From 1917, when the city

The Midway, with the Grandstand beyond, as it appeared during the 1936 fair (Courtesy Daily Oklahoman).

purchased the fair grounds, to 1937, the darkest depths of the Great Depression, the State Fair of Oklahoma faced a succession of challenges that remained largely hidden from the general public.

First came a change of leadership. I.S. "Dick" Mahan, hired as secretary and general manager of the fair in the winter of 1907-1908, had successfully guided the fair through the turbulent transition to city ownership and helped mold the program mix and image of the fair. Nationally, he had been just as successful, assuming a leadership role in the International Association of Fairs and Expositions, where he earned a reputation as a knowledgeable "fair man." His abilities were good; his health was not. In April of 1922 the fair board granted him a six-month leave of absence due to illness. In February the leave was extended. In April of 1923 he died.[48]

His successor was Ralph T. Hemphill, who would guide the fair for the next 35 years. A native of rural Illinois, Hemphill came to Oklahoma in 1909 and soon became a county extension agent. He served as assistant director of extension at Oklahoma A&M from 1915 to 1917 when he was hired by Mahan as livestock manager of the State Fair of Oklahoma. By the time of Mahan's death, Hemphill was secretary of the fair board and a

virtual equal with Mahan in managing the day-to-day operations of the fair.[49]

Hemphill, according to contemporary accounts, was both respected and liked. He worked well with the board and developed close ties with businessmen and politicians. Professionally, he followed Mahan's lead and became active in the International Association of Fairs and Expositions, serving that organization as president from 1925 to 1930. He also served as a director of the International Motor Contest Association, an early-day sanctioning body for automobile racing. To Hemphill and his board fell the task of coping with the changing world of running a state fair.[50]

One of the first challenges was also one of the most bizarre. Governor John "Jack" Walton, who had been mayor of Oklahoma City in 1919 and 1920, was locked in a bitter battle with the State Legislature as the 1923 fair approached. Walton, to prevent the legislature from meeting and therefore impeaching him, declared martial law in Oklahoma City and called out the National Guard. In an act of desperation, he even threatened to close the fairgrounds, stating that he would allow it to open only if E.K. Gaylord and the Oklahoma City press ceased the critical reporting of his actions. Despite the threat the fair opened on schedule, and Walton was ultimately impeached.[51]

A more serious threat was the relentless cost of maintaining the grounds and buildings. In February of 1926, Hemphill publicly announced that the fair association could not "replace the rapid depreciation" of the older buildings. Even the buildings constructed with bond money in 1918 and 1919 were showing wear. Faced with

Floyd S. Lamb was president of the fair board in 1935 and 1936, just as the Great Depression and oil exploration were taking a toll on the State Fair of Oklahoma (State Fair Archives).

Left and opposite page: The Midway in 1934 was lined with barkers and "teasers" of the "Big Show Inside" (State Fair Archives).

mounting expenses and little support from the city, Hemphill suggested to the board that it approach the Oklahoma City Chamber of Commerce and establish a closer long-term association to generate support for the fair.[52]

There was good reason for the strategy. The Chamber had always supported the fair, a natural alliance forged as early as 1892 on the twin anvils of economic development and town boosterism. In 1924, in an expression of renewal, the new manager of the Chamber addressed the fair board: "The Chamber of Commerce" he said, "pledges its support towards making the Made in Oklahoma Exhibit a success....We also are ready and willing to cooperate in every way possible with the officers and directors of the Oklahoma State Fair in making the 1925 fair a success." The new manager making that commitment was Stanley Draper, a tireless worker who was just starting a remarkable career as city builder.[53]

Fortunately for the fair, Draper had the full support of Ed Overholser, president of the Chamber of Commerce and the son of Henry Overholser. He also had been mayor when the city purchased the fairgrounds and passed the bond issue in 1917, so he understood the importance of the public-private partnership. That commitment was quickly tested in 1926 when five days of continuous rainfall plunged the fair deeply into debt.

Overholser and Draper reacted quickly to the crisis by raising $37,000, enough to pay the debt on the 1926 fair. Still, there was no money for operational expenses and no prospect for solving the long-term maintenance problems. Faced with bankruptcy, the fair board asked the community to raise $25,000, or the "institution would be closed."

The Chamber stepped forward again. Twenty-five businessmen guaranteed payment for a $25,000 loan, then applied pressure to the city council to appropriate money for maintenance. In September of 1927, just

Food vendors typically arrived at the fairgrounds a few days before the fair began so they could construct temporary booths from wood and canvas. These nomadic merchants offered chili, hamburgers, hot coffee, cold drinks, and a "plate lunch for 25 cents" (Courtesy Daily Oklahoman).

The Great Depression, when coupled with the problems of oil production and lack of parking space, pushed the State Fair of Oklahoma into a period of belt tightening and smaller operations. Here, federal relief workers get tools to improve Brock Park in Oklahoma City (Courtesy Oklahoma Department of Libraries).

before the fair opened, the city allocated $25,000 for roof repairs, walks, roads, and fences.[54]

With this financial breathing room, the fair board and city worked out a contractual arrangement that transferred all maintenance responsibilities to the fair board. In return, the fair could keep all revenues from the rent of buildings and facilities throughout the year. With horse shows in the spring, dog and horse races during the summer, and rent of storage space to local businesses, the extra income bought a few more years for the cash-strapped association. By 1930 the budget for the fair was $146,000, making it one of the top five fairs in the nation.[55]

This expanded budget in 1930 seemed reasonable despite the crash on Wall Street in October of 1929. The engines of the local economy roared with vitality, fueled by increased production in the Oklahoma City Oil Field and paced by a construction boom that included the Nichols Hills Country Club addition and a booming skyline pierced by the 31-story First National Bank Building, the Ramsey Tower, and the Biltmore Hotel. That era of prosperity, unfortunately, was about to end.

The Great Depression hit the state and city with blinding force. Buildings such as the Skirvin Tower stood unfinished. Companies laid off thousands of workers. Houses sat vacant as people simply moved away, unable to make mortgage payments. By 1931 the Emergency Employment Committee reported that more than 4,200 families in Oklahoma City needed immediate financial support just for survival. Many lived in "Hoovervilles," squatter camps that appeared around dumps, along river banks, and beside highways. The State Fair of Oklahoma could not avoid the impact of such economic disaster.

The fair board gradually cut back its budget, dropping events such as the horse show and auto racing. By

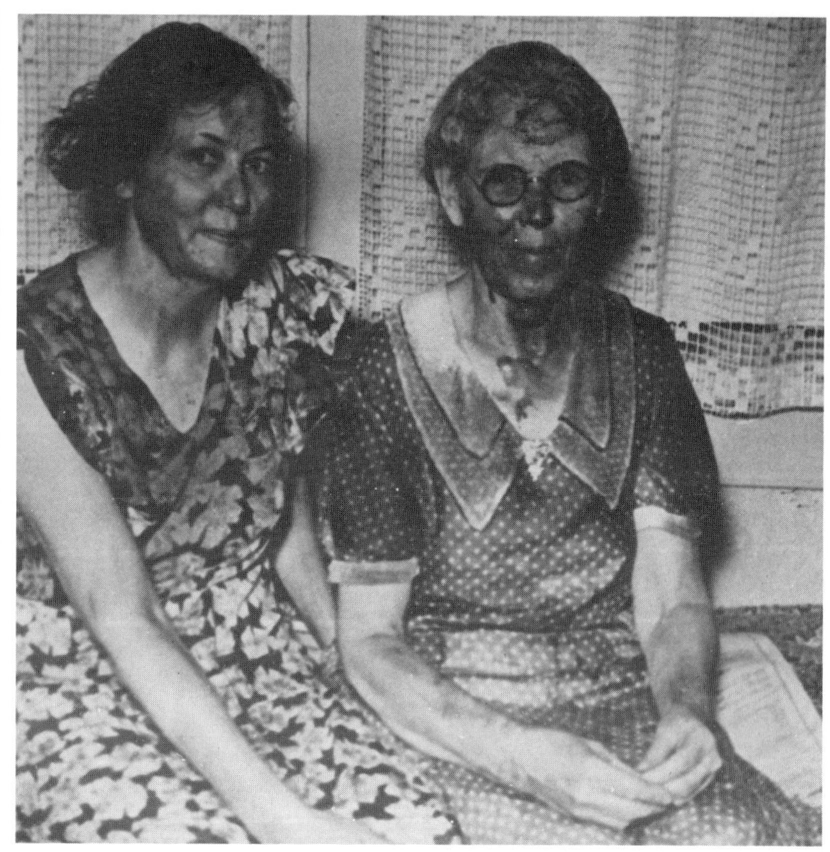

Another consequence of oil production in a major city is seen on the faces of these women, who were covered with oil from a blowout on the east side of town (Courtesy Daily Oklahoman).

1933 the budget was down to $63,000, less than half of what it had been three years earlier. Still, it was not enough to make ends meet. Deficits grew, from $11,629 in 1931 to $26,528 in 1932. The crisis reached a low point in June of 1933 when the First National Bank, a longtime ally of the fair, refused to make further loans for operating expenses. The losses had been too great and the prospects for recovery did not look good.[56]

As before, the people of Oklahoma City rallied to the cause. A brewery loaned the board $10,000 in exchange for exclusive rights to wholesale beer during the fair, and the city appropriated $25,000 for premiums. The Chamber, in its growing role as guardian angel, helped raise emergency funds as needed. Pulling together, the fair survived the immediate crisis and broke even the next year, although at a reduced level of expenditures.[57]

The financial woes only drew attention to more fundamental problems facing the fair. Foremost was a lack of space. For several years Hemphill had complained that the fairgrounds were too small, and the situation had grown worse. Since 1907 almost 40 acres of the quarter section had been unusable owing to regular flooding of the river and the surrounding wetlands on the east side of the grounds. Then came the demands of parking space as Oklahomans joined the "automobile age." The fair board built the first on-site parking lot in 1920, but the allocation of space for automobiles never seemed to be enough. Hemmed in by the river on the south and residential neighborhoods on three sides, there was simply no room for expansion.

The space problem, previously seen as a simple burden, became a full-blown crisis in the 1930s. The problem, ironically, was oil. The Oklahoma City Oil Field had spread north from the discovery well at S.E. 59th and Bryant until it skipped the river and invaded developed neighborhoods on the east side of town. With city permission, the first well was drilled on the fairgrounds in 1934. It produced more than 100 barrels an hour, so two more wells were drilled in 1935. With rigs, tanks, machinery, and all the attendant noise and congestion of exploration and production, acreage on the fairgrounds suitable for the fair grew smaller and smaller. By September of 1936, headlines in the local newspaper recognized what the fair board had been predicting, "Oil Wells Doom Present Location of Fair Park."[58]

The campaign to move the fair gained widespread

As seen in this aerial shot of the old fairgrounds, all available space had been utilized either for exhibits or parking by the late 1930s. As more people drove automobiles and depended less on public transportation, the parking problem convinced many people in Oklahoma City that a new, larger location for the fair was necessary (Courtesy Daily Oklahoman*).*

support as other issues were addressed. Some city leaders wanted to host a semicentennial celebration of the land run in 1939, and to do that they needed a larger fairgrounds. Others, seeking more revenue for the city, wanted to let the oil play spread unopposed across the existing fairgrounds. Still others saw a solution to the need for more housing in the black section of town. "Elimination of the fair park," wrote one observer, "would pull the cork from the bottled-up Negro district and allow it to expand to the east."[59]

Several sites were suggested for a new fairgrounds. Edgemere Park, nestled on the near north side between Robinson Street and Walker Avenue, was considered but rejected as too small with surrounding land too expensive. The favorite of most supporters was Southwest Park, located at S.W. 29th and May Avenue, a tract that had been set aside from development in 1909 as part of the Grand Boulevard parks system. The site met all of the criteria. It was served by railroads and highways; it was near the new municipal airport; and most importantly, it was available. To guarantee that enough land would be set aside, the fair association took out an option on an adjoining quarter section.[60]

With typical aggressiveness, Hemphill, the fair board, and the Chamber of Commerce assembled a package that combined city, state, and federal resources. They lobbied the state legislature and succeeded in passing a bill providing $320,000 for a new fairgrounds. The only catch was that they had to come up with $2 million in matching monies. Hemphill quickly went to Washington where he failed to get a WPA grant, but did get a commitment from Senator Elmer Thomas to sponsor an appropriation of $1 million for the fair. Then the Chamber organized a Citizens Advisory Committee and proposed a $600,000 bond issue to "move the fair." Everything seemed to be coming together for a new and bigger State Fair of Oklahoma. All they needed was the consent of the city council to put a bond issue before a vote of the people.[61]

A majority of city council members, despite the potential funding from state and federal sources, decided that they wanted no part in funding an expanded fair. Not surprisingly, the most vocal opposition to the bond issue came from the two councilmen in the northeast ward, who were "pledged to resist removal of the fairground from the east side." While they cited financial reasons for their opposition, the editorial writer of the *Daily Oklahoman* detected another motive. "In the opinion of councilmen Moore and Jones," he wrote, "it would be better to have an empty and abandoned fairgrounds on the east side than to run the risk that if it were moved, the location it now occupies might conceivably fall into

Local leaders successfully coupled city funds, voter support, and federal grants to build the County Courthouse from 1935 to 1937. A similar effort for a new fairgrounds failed due to local politics (Courtesy Oklahoma Historical Society).

the hands of Negroes and be used either for a Negro park or for Negro residences." By a vote of 5 to 4, the city council rejected the proposal for a bond issue, and thereby rejected the state and federal matching funds.[62]

Defeat with victory so close must have been devastating to Hemphill, the fair board, and the Chamber. After all, they were locked in the grips of the Great Depression with a fair half the size of its former self. Space for the fairgrounds was contracting as oil wells and tanks replaced park lands and rides. Adding insult to injury, not one penny of oil revenue was earmarked for the fair. An era that had begun with such high hope—the city purchase and the bond issue—was ending in dark despair. Once again, the leaders of the State Fair of Oklahoma were facing a crisis. Once again, the future of the fair was in the balance.

Chapter Five

THE BIG MOVE

Defeat of plans to move the fairgrounds in 1937 was a setback but not the death blow some people had predicted. As before, supporters rallied behind the fair and the show went on, albeit in a limited way with few prospects for improvement.

Although the Great Depression dragged on, the fair slowly recovered from its low point in 1932 and 1933. Each year attendance rebounded a little more, and revenues allowed the fair board at least to break even. In terms of programs, there was little expansion, just a continuation of traditional competitions and entertainment features. One of the few innovations during this troubled era was the addition of women competitors to the rodeo, which gained in popularity throughout the late 1930s and early 1940s.

World War II and the Japanese attack on Pearl Harbor swept away any lingering complacency. The biggest shock was the widespread opinion that all fairs should be cancelled so the nation could more effectively prepare for war. Hemphill and other fair directors quickly replied that fairs could help, not hinder, preparedness by galvanizing and focusing public support. As if to prove their point, fair directors across the nation used soldiers, sailors, marines and their equipment to lure curious crowds and educate the public about the challenges ahead.

Opposite page: The State Fair of Oklahoma, as it looked in 1957, only three years after the "big move" (State Fair Archives).

War was the dominant theme at the 1942 State Fair of Oklahoma. "Preparation Day," on September 26, 1942, featured displays of guns, bombs, uniformed men, pup tents, airplane engines, and all the accoutrements of war except a battle front. The Will Rogers Field Army Air Corp Bomber Base brought in a P-40 Fighter Plane and set up a parachute packing demonstration. Homer C. Challios, Indianapolis American Legion Director of Americanism, was the featured speaker at the Army-Navy Day.[1]

As the fighting escalated, the military opened a string of bases, depots, and prisoner of war camps across the state. Personnel and exhibits from these installations naturally found their way to the fair. The biggest attraction at the 1944 State Fair was a battle-scarred B-17, one of the famous "Flying Fortresses," that had flown 118 missions in the South Pacific. Strangely, it had never acquired a name although it was credited with the destruction of 20 Zeros and the sinking of six Japanese ships. The public could walk through the plane and see machine guns, armaments, and compartments. The plane was one of 100 at Tinker Field awaiting repairs and return to active service. The B-17 exhibit was credited with record attendance at the 1944 fair.[2]

The war effort, despite the best efforts of Hemphill and the fair board, still caused disruptions. Oklahoma City University operated an Air Corps Cadet Training Program on the fairgrounds for two years, occupying

The war effort, illustrated here by B-24s under construction at Tinker Air Force Base in 1943, dominated the fairs from 1942 to 1944 (Courtesy Tinker Air Force Base).

both the 4-H Building and the FFA Building. During the fair the 4-H boys and girls were housed downtown in the Municipal Auditorium, while the FFA boys camped in a large tent on the fairgrounds provided by the State Fair. There also were shortages of some supplies, such as sugar and other foodstuffs, that had been regular staples at the fair.[3]

Despite adjustments and shortages, the fair posted steady surpluses during the war, the result of increased attendance, an economic boom due to plentiful jobs and full production, and a deferment of capital improvements as building materials went into the war effort. In stark contrast to the 1930s, the fair board was able to purchase $50,000 in government bonds from their reserves. The downside was a gradual deterioration of the physical plant. Hemphill tried several measures to improve facilities, including a request for funding from the State Legislature, appropriations from the city, and borrowing funds to make the improvements. All efforts failed. Instead, he concentrated on minor improvements in the Livestock Pavilion and put his proposal for building a new dairy barn on the shelf.[4]

The earlier threat to cancel all fairs finally became reality on June 8, 1945, when the United States Government War Committee on Conventions ordered the 1945

Sweet treats such as this wafer sandwich became a rarity during the war (Courtesy Daily Oklahoman).

Even with restrictions on automobile travel during World War II, cars overflowed the limited parking available at the old fairgrounds (Courtesy Daily Oklahoman).

Opposite page: Television, first seen in Oklahoma in 1949, would pose a new challenge to the fair as the "Baby Boomer" generation emerged after World War II (Courtesy Daily Oklahoman).

State Fair cancelled. Frank Perrin, secretary of the committee, stated that the closing was necessary to reserve all public transportation facilities for the movement of men and material for the war effort. Accordingly, the fair board authorized Hemphill to cancel all contracts for the 39th Annual Oklahoma State Fair and Exposition for the dates September 22 to 28, 1945. To make matters worse, there was a disastrous fire that summer.[5]

Faced with the moratorium, Hemphill suggested and the fair board agreed to host a local festival instead. The traditional agricultural displays were cancelled, so Hemphill scheduled the Oklahoma County Free Fair for a three-day run during the festival, which according to one reporter, "provided some agricultural flavor." Obviously protecting himself and the fair board from federal intervention, Hemphill billed the festival as the "biggest strictly entertainment event ever staged in Oklahoma."[6]

There was no admission charge to get onto the grounds, but people did pay to see special events. Attractions included the Barnes and Carruthers "spectacular grandstand show" every night and the Royal American Midway as usual. Another feature repeated from the past was the automobile thrill show, which, according to press releases, "proves that it's possible to wreck automobiles, ride through brick walls, and jump over a bus lengthwise without getting killed." Still, the festival lost money, forcing the board to cut all staff salaries the following year. It also was proof that the traditional appeal of agricultural competition and displays were critical to the success of the State Fair of Oklahoma.

The financial losses of the 1945 fair, when added to the progressive deterioration of the facilities, fueled new doubts about the future of the fair. After the war ended and people once again looked to the future of the city and state, the burning question once more arose as to whether or not the fair should be moved to a new location or expanded at the existing site.[7]

Problems with the site on Eastern Avenue were obvious. It was too small, with less than 160 acres, to accommodate both the fair and the growing demand for more parking. An additional problem was oil production on the site, reducing usable space and posing a noisy, messy intrusion in what was meant to be a pastoral park. Too, the facilities were simply worn out, the result of long-delayed improvements caused first by the Great Depression, then by the war. With the fighting over, the question could no longer be delayed. The fair board had to either invest vast sums in the old grounds or move to a new site and start over again.

That question—whether or not to move—was just one of the many issues facing the leaders of Oklahoma City after the war. As men like Henry Overholser and Stanley Draper had always known, major decisions for institutions such as the fair had to be made against the back-

Below: The 1950s was a decade of change not only for the fair but also for Oklahoma City, seen here looking from the northeast to the southwest (Courtesy the author).

drop of the bigger picture, in this case the postwar development of Oklahoma City's economy. If the fair was to grow and continue to be an important part of community life, it had to fit within the framework of the city's future. And that future in large part was being shaped by the Oklahoma City Chamber of Commerce.

The Chamber had always had a strong influence on both the city and fair. The territorial fairs had been

Hereford judging at the 1941 State Fair of Oklahoma, inside the Livestock Pavilion, the only structure that would be moved to the new location 13 years later (Courtesy Daily Oklahoman*).*

organized by a committee of the Chamber, and the first meeting to organize the state fair in 1907 had been held in Chamber offices. In 1917, when the fair needed city support to avoid bankruptcy, the Chamber had successfully lobbied the voters for a bond issue and commitment of public resources. That helping hand had been offered consistently in the 1920s and 1930s every time a fund-raising drive was needed. The Chamber's support had been neither an accident nor an act of charity; it was the result of calculated self-interest, an ongoing affirmation that the fair was an important and essential piece in the puzzle of economic development. In the post-World War II era, this link between the fair and general prosperity was stronger than ever.

Just as the postwar economic boom began, the Chamber established a full-time industrial division to solicit new industry, give assistance to established industry, and create a better business climate. The Chamber also organized a development corporation called Oklahoma Industries, Inc., dedicated to the development of industrial parks and construction of industrial buildings for

Circus acts were staged in the Grandstand infield. To the left, in the foreground, is the Livestock Pavilion (*Courtesy* Daily Oklahoman).

lease or sale. By 1966, as a public trust with the power to issue revenue bonds, the authority had financed a wide range of facilities for health care and education in addition to the traditional role of industrial facility financing. By the 1990s the Oklahoma Industries Authority would be the largest in the Southwest and one of the largest of its kind in the nation.[8]

Another of the Chamber's traditional priorities that expanded in the postwar era was the development of aerospace-aviation facilities. Through the Industries Foundation in the early years of World War II, the Chamber had been instrumental in developing Tinker Air Force Base, the Bomber Base at Will Rogers Airport, and both Wiley Post Airport and Clarence Page Airport on the west side of Oklahoma City. Following the war, the initiative continued through another entity of the Chamber, Greater Oklahoma City, Inc., which purchased more than 18,000 acres around Tinker and 5,000 acres around Will Rogers Field to protect both from encroachment and allow for future growth. By the 1990s total employment at the combined facilities of these aerospace-aviation complexes represented about 20 percent of all the work force in the greater Oklahoma City area.[9]

Another initiative affecting decisions surrounding the future of the fair was the urban expressway system of Oklahoma City. Early development of the system was accomplished by the Chamber when it borrowed funds to acquire right-of-way, then sold the acreage at cost back to the appropriate governmental agencies. To accomplish this, it was essential to secure the cooperation of the City of Oklahoma City, Oklahoma County, the State Department of Transportation, the Oklahoma Turnpike Authority, and the Federal Highway Administration. The Chamber retained Treat Engineering Company in 1946 to do the broad planning for a city-wide, integrated transportation system.[10]

Guided by Stanley Draper, the Chamber since the 1920s had achieved remarkable results by serving as a coordinating body that could move quickly and efficiently as a private organization and pull together the major forces of public policy for the larger good of the community. It was in this role of "deal maker" that Stanley Draper and the Chamber staff put together one more initiative that included the future of the State Fair of Oklahoma.[11]

The deal in this instance involved the acquisition of a section of school land located on the west side of Okla-

With gasoline rationing over, car races returned to the fair in 1947 (Courtesy Daily Oklahoman).

Large crowds watched car jumping at the 1947 fair (Courtesy Daily Oklahoman).

homa City, a 640-acre site bounded on the north by N.W. 10th Street, on the east by May Avenue, on the south by Reno Avenue, and on the west by Portland Avenue. The Chamber, always eager to promote Oklahoma City as a regional center for farmers and ranchers, proposed the location of a demonstration farm on the northwest quarter. On the northeast corner, Draper wanted a technical high school which could train new workers for the coming industrial boom that he and the Chamber predicted. On the southeast and southwest quarters, which would be at the junction of two future express highways, Draper envisioned the perfect 320-acre site for an expanded and improved State Fair of Oklahoma.

Typical of Draper's method of operation, the Chamber purchased preference rights held by the various leaseholders on the school land, then borrowed funds guaranteed by Chamber members to back his purchases. Theoretically, the Chamber would eventually assign its preference rights at cost to the ultimate purchasers, in this scenario, Oklahoma A & M College, Oklahoma City Public Schools, and the City of Oklahoma City, custodian of the Oklahoma City fairgrounds.

A&M College proceeded quickly with its part of the master plan to develop a demonstration farm. A special act of the legislature was required for it to purchase the

Louis A. Macklanburg, president of the fair board from 1952 to 1955, worked with Pete Baker in developing the new fairgrounds and getting everything ready for the grand opening in 1954 (State Fair Archives).

quarter section, construct farm buildings, terrace the land, plant crops, and purchase livestock. Fortunately, Henry G. Bennett, longtime president of A&M, was at the peak of his political power and personally managed the legislation that gave the college authority to purchase the property as well as an appropriation to fund the project. This legislation sailed through both houses of the legislature without opposition.

The concept of a demonstration farm was short-lived, however, and never reached its expected potential. Ultimately, the land for the project would be designated as the campus of OSU Technical Institute, later the Oklahoma City Branch of Oklahoma State University. In a broad sense it fulfilled the concept of a technical high school, but at a higher level it was another important cog in the industrial development of Oklahoma City.

Draper did not encounter the same cooperative spirit with the fair board, especially its secretary-manager, Ralph Hemphill. The battle lines were drawn by Hemphill at the annual meeting on November 12, 1946, when he reported it was the sense of the shareholders that the fairgrounds should be retained and enlarged at its longtime location on Eastern Avenue. To Draper's dismay, the motion was adopted.[12]

Hemphill also moved that the members recommend to the fair board that the livestock pavilion be rebuilt at the earliest possible time. Again, the motion was adopted. Another motion was approved to ask the fair board to get an appropriation from the state legislature and initiate a bond issue that could be submitted to a vote of the people at the April 1947 city election. The bond monies would be used to rebuild the fairgrounds at the old location with a new sports stadium. As a result of that one aggressive meeting, Draper's plan for a new fairgrounds seemingly had reached a dead end.[13]

The setback was only temporary. The change of heart started in 1947 when two new members were elected to the fair board. One was Ancel Earp, well-known insurance executive, and the other was Louis Macklanburg, chief executive officer of the Macklanburg Duncan Company; both men were close associates of Stanley Draper. Then on November 11, 1947, Tom Dee, general manager of Armour & Company, at that time one of Oklahoma's largest meatpacking plants, was elected president of the fair board. Dee was another of Draper's allies.

At a fair board meeting on September 13, 1948, Dee reported that he had talked with Hemphill's doctor, Hugh Jeter, who said that Hemphill's health had deteriorated to the point that he should not continue his work. The board granted Hemphill a leave of absence and relieved him of all duties until the annual meeting on November 11, 1948.[14]

Ancel Earp played a major role in convincing the fair board to move to the west side location (Courtesy Daily Oklahoman*).*

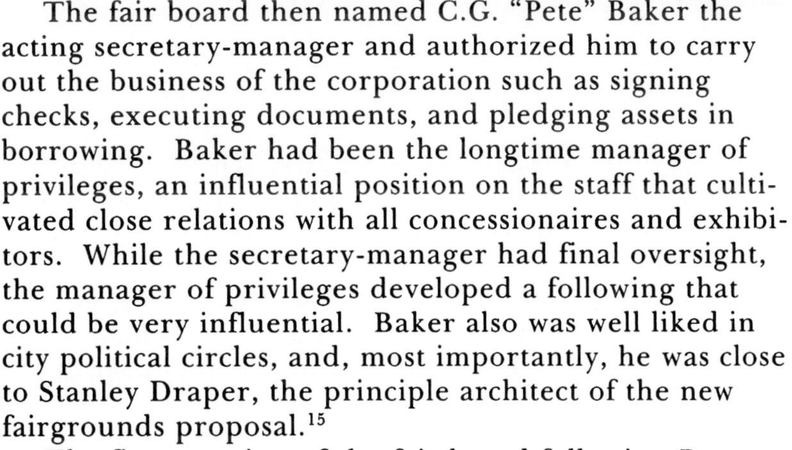

After years of neglect, then abandonment, the old buildings at the original fairgrounds were either moved or razed (Courtesy Daily Oklahoman*).*

The fair board then named C.G. "Pete" Baker the acting secretary-manager and authorized him to carry out the business of the corporation such as signing checks, executing documents, and pledging assets in borrowing. Baker had been the longtime manager of privileges, an influential position on the staff that cultivated close relations with all concessionaires and exhibitors. While the secretary-manager had final oversight, the manager of privileges developed a following that could be very influential. Baker also was well liked in city political circles, and, most importantly, he was close to Stanley Draper, the principle architect of the new fairgrounds proposal.[15]

The first meeting of the fair board following Pete Baker's promotion was pivotal. Baker recommended that the fair board appoint a member to serve on the Chamber's planning committee for a new fairgrounds. Serving on the committee were representatives of the

school board, park board, city planning commission, and the Chamber of Commerce. The fair board chose L.A. Macklanburg to represent the state fair. Stanley Draper then was called upon to outline plans for a new fairgrounds. He estimated the cost at $15 million and recommended a general obligation bond issue to finance the project.[16]

Meanwhile, the Oklahoma City School District moved ahead with its plans to build a technical high school on the northeast quarter of the west side section. The district purchased its quarter section from the Chamber but then ran into trouble that ultimately doomed the project. Instead, it was decided that the district needed a new high school for black students, most of whom were concentrated in segregated neighborhoods on the northeast side of town.

To Draper, the solution seemed obvious. The school board, left with a quarter section of land on the far west

side for which they it no use, could trade that parcel of land to the city. In exchange, the city could give the school district the old fairgrounds at N.E. 10th and Eastern, the perfect location for a new black high school. When it was determined that a straight exchange was not legal, the school board and city council executed a sale-purchase plan. The city thus acquired the northeast quarter on May Avenue and sold the school board 40 acres of the old fairgrounds. Within a few years, Douglas High School was built on that land. With land and a conscious decision to move to the west side location, the fair board turned to internal reorganization.

At the annual meeting held on March 9, 1949, shareholding members adopted new bylaws, eliminated capital stock, and created a voluntary association with membership certificates in lieu of stock. One year later, the board officially elected Pete Baker as secretary-manager

The same section of land, seen here in 1993, after the fair was developed at the May Avenue site. Interstate 240 was built along the old course of Grand Boulevard, dividing the section between the fair (on the right), Oklahoma State University Tech (lower left), and Oklahoma City training facilities (top left) (Courtesy Ace Aerial Photo Service).

and awarded Ralph Hemphill a $500-per-month retirement for the ensuing year. Ancel Earp was elected president and L.A. Macklanburg became first vice-president. Lem Jones was elected superintendent of privileges and auditor to replace Pete Baker.[17]

With the leadership team assembled and unified, the fair board gave full support to the campaign for a new fairgrounds. The Chamber of Commerce, meanwhile, had organized the "Greater Oklahoma City Committee"

The General Exhibits Building, "designed for modern commercial exhibits and trade shows." A corridor joined it with the Women's Building (From "Panorama of a Dream").

to promote a city-wide bond issue of $35,974,000. Of that amount, $4,750,000 was designated for the fair, well under the proposed estimate of $15 million needed to build a completely new fairgrounds.

Despite the inadequate level of funding, support among the staff, board, and general public rallied behind the need for a new fairgrounds. The Chamber retained the Treat Engineering Co. of Oklahoma City to prepare a preliminary master plan at a cost $20,000. The fair board pledged one-third of the cost to be expended prior to the vote on the bond issue and authorized payment of one-seventh of the total campaign fund.[18]

The bond issue covered a wide range of improvements that touched all residents of Oklahoma City. Projects included: $600,000 for airports; $4,750,000 for the fairgrounds; $600,000 for the fire department; $2,775,000 for floodway control; $822,000 for libraries; $2,100,000 for parks; $329,000 for the police department; $4,540,000 for sanitary sewers; $2,710,000 for storm sewers; $6,500,000 for streets; $200,000 for traffic control; and $10,048,000 for the water department. To market the proposal, the Chamber adopted a logo of an ostrich in a small cage with his neck protruding out of the cage with a knot tied in it. They used the byline "A pain in the neck." The bond issue, the first major capital investment proposed after World War II, was passed by voters on May 9, 1950.[19]

The next task facing Baker and the fair board was just as daunting—how to spend the $4,750,000 for a new fairgrounds. The three major entities responsible for the project met on May 22, 1950. Representing the city were Mayor Allen Street, City Manager Ross Taylor, and Councilmen Errett Newby, Laverne Carlton, Walter Harrison, Warren D. Conner, Harlou Gers, and G.A. Stark. The Park Board was represented by Superintendent Pat Murphy and Park Board members George Smith, Floyd Broadersure, and J. Herndon Lackey. Representing the fair board were members Frank Martin, Ancel Earp, L.A. Macklanburg, Tom Dee, R.L. Peebly, J.M. Owen, and F.S. Lambe. Also attending were Pete Baker and Stanley Draper.

The most important decision made at that pivotal meeting was the selection of Phil Wilbur, Dean of the School of Architecture at Oklahoma A&M College, to develop a master plan and building design for the fairgrounds. Mayor Street then appointed councilmen Laverne Carlton, G.A. Stark, and Errett Newby to consult with the fair board in drawing up a new operating contract. Finally, it was the consensus of the mayor and council that the fair board should recommend architects and engineers for each of the new buildings within the context of the master plan.[20]

Judge Robert M. Rainey served the fair as president from 1939 to 1941, years of recovery for the State Fair of Oklahoma (Courtesy Daily Oklahoman*).*

The Agriculture Building, designed for the display of "agriculture exhibits, poultry and dairy products" (From "Panorama of a Dream").

With this direction, the executive committee of the fair board recommended that Oklahoma A&M College architects and engineers be employed as consultants for the new fair plant at a fee not to exceed one-half of one percent of the amount allocated for the new fairgrounds. They further recommended that each of the new buildings be assigned to a different architectural firm. The list included Park & Adderhold for the Arena, Wright & Selby for the Grandstand, Sorey, Hill & Sorey for the agricultural buildings, Coston & Frankfort for the exposition facilities, Nofsgar & Lawrence for livestock buildings, and Walter Valberg for service facilities. Guy Treat was recommended for all engineering needs.[21]

Although long-range planning quickly came to dominate meetings of the fair board, running the annual fall fair still demanded attention. The fair continued at the old fairgrounds for the next four years, even though facilities continued to deteriorate. With minimal investments, operations became more profitable. The surplus following the 1950 State Fair was in the range of $160,000.[22]

Composition of the board also continued to change. At the annual meeting of 1950 two longtime members of the fair board retired owing to poor health. They were Oscar Deitz and J.M. Owen. Owen had served as president in 1917 and 1918 and again in 1924. As president in 1917, his name appeared on the first agreement and contract between the city and the fair board. Owen's son,

The Appliances Building, intended to host "trade shows, sporting events, an ice rink, and other purposes" (From "Panorama of a Dream").

The Oklahoma Publishing Company Building, the "first of many permanent, private buildings to be constructed on the fairgrounds" (From "Panorama of a Dream").

Nelson, would later serve as a director from 1951 to 1972.[23]

Personnel matters also remained an important issue. When Ralph Hemphill retired in 1950, he was granted a monthly salary of $500, for there was no retirement plan. Other retiring employees had been and would continue to be granted monthly stipends even though there were no provisions for funding. That practice would ultimately have a negative impact on operating funds. At the annual meeting in 1950, members agreed that a study was necessary to find a way to implement an employee retirement plan, an objective that would not be accomplished for another decade.[24]

Meanwhile, other groups bombarded the fair board with proposals for use of the new fairgrounds. The Oklahoma Fish & Game Department, supported by Governor Johnston Murray, proposed the building of a large aquarium and office building. The Oklahoma Farm Bureau expressed interest in putting up a building to house their offices and other agriculture organizations on the fairgrounds. Inexplicably, the fair board approved a recommendation to use 150 acres of the site for an airport. There were a number of other issues proposed, and later some were approved.[25]

In August of 1951 the City of Oklahoma City, Oklahoma A&M College, and the Board of Education made the final adjustments to the section of land at 10th and May Avenue through a series of purchases and sales of various parcels. In addition to these transactions involving both the old and new fairgrounds, the county purchased right-of-way for Interstate 44 with connection to

Tom J. Dee served on the fair board from 1944 to 1957, two years as president, spanning the move from the old location to the west side (State Fair Archives).

*George A. Davis was president of the
fair board from 1942 to 1944, a
trying period for the fair due to
shortages and then the cancellation of
the 1945 fair (State Fair Archives).*

Interstate 40. This more or less set the development
boundaries of the fairgrounds share of the new section.
In the final analysis, Oklahoma A&M College ended up
with approximately 200 acres, while the City had approximately 440 acres on which the new fairgrounds
would be established. In relation to future landmarks, all
land east of I-44 would end up as part of the fairgrounds;
the land on the west side would be used for the Oklahoma City branch of Oklahoma State University and the
Police & Firefighters Academy of Oklahoma City.

The location of railroad right-of-way on the section
was another problem that needed to be resolved. The
Chamber had estimated it would cost $1 million to re-
move the tracks and right-of-way, but the actual cost ran
to $1,206,000. When the Chamber asked the fair board
to share part of this cost, the board voted to advance
$50,000. Through a series of negotiations it was deter-
mined that it was only feasible to remove the old line
running through the north half of the section. The
existing main line would remain.

In 1951 members of the fair board elected Louis A.
Macklanburg president. This was an important and
fortunate move, for Macklanburg would serve during the
next three years as the new fairgrounds was being constructed. His expertise as a successful businessman was
just what the fair needed at that particular time in history, and he was a close friend of Stanley Draper. They
worked well together, and Macklanburg was able to
control Draper's aggressiveness. The combination of
Macklanburg and Draper accomplished much for the
State Fair of Oklahoma.[26]

Ground-breaking ceremonies were set for September 1, 1953, under the auspices of the Chamber of Commerce. The Chamber named Major General Roy W. Kenny, State Adjunct General, as chairman of the event and James Burge, special events director of Oklahoma Publishing Company, as vice-chairman. More than 3,000 people attended the festivities, which included band concerts, a search for 410 gold pieces, and fireworks. The first shovels of dirt were turned by State Fair President Lou Macklanburg, acting mayor Leonard Dickerson, and Chamber President William Morgan Cain.[27]

Louis Macklanburg, president of the fair board, took the controls for a publicity shot as work began on the new fairgrounds (State Fair Archives).

The fairgrounds quickly took on the look of a combat zone. Men and equipment descended on the site and started grading building sites, laying out streets, and putting in utilities. The most daunting individual challenge was construction of the Grandstand, 180 feet wide by 360 feet long, designed to be the most modern in the nation. It featured a large stage and dressing rooms in the infield, two dirt tracks, seating for 12,000, and offices and ticketing facilities.[28]

The biggest enclosed space on the new fairgrounds was the General Exhibits Building, measuring 350 feet long by 105 feet wide. Designed for quick expansion into an "L" shape, it was to be the "central showplace of the new fair plant" with a distinctive exterior described as "saw tooth brick and steel construction with exposed steel

Opposite page: Balloonist Garrett Cashman made a brief but unusual flight at the 1955 fair (Courtesy Daily Oklahoman).

Vic Hyde, announcer for the "Fair Follies" in 1947, displayed his talent with three trumpets (Courtesy Daily Oklahoman).

frame." It was connected by an enclosed corridor to the Women's Building, which was smaller, with three bays "particularly suited for cooking schools, style shows, and events requiring stage and exhibit facilities."[29]

Nearby was the Appliances Building, designed for trade shows, sporting events, and an ice rink. With an inside area 120 by 240 feet clear of posts, it had a distinctive appearance created by the curved arches of the roof line and the "colorful modern architectural design of the new fair plant." The modernistic style also was incorporated into the Oklahoma Publishing Company Building, with its landmark tower offering visitors a bird's-eye view of the entire fairgrounds.[30]

On the north side of the central plaza was the Agriculture Building, billed as the "focal point of the fair, where agriculture exhibits, poultry and dairy products will be displayed." It was 372 feet long by 60 feet wide, with low sweeping gables creating a look reminiscent of Oklahoma's rolling plains. Nearby was the Livestock Pavilion, the 1912 structure dismantled at the old site and moved to the new location, and three more steel frame barns designed for stalls, auctions, and competition.[31]

Anchoring the new fairgrounds on the east were the Future Farmers of America Building and the 4-H Club Center. The FFA exhibit hall was 248 feet long by 82 feet wide with an "L" extension 70 by 80 feet. It included dormitory space for 1,000 boys, exhibit areas, and a modern cafeteria. The 4-H Center also included space for dormitories, an exhibit hall, and a cafeteria that could serve 1,500 persons at each mealtime.[32]

Typical of such ambitious construction projects, the ultimate challenge was not design but time, as workers rushed to meet the deadline of September 24, 1954. One of the most serious obstacles was the weather. The summer of 1954 was one of the hottest and driest on record. On July 12 the temperature rose to 105. A construction worker on the fairgrounds, C.T. Overton, collapsed and died.[21] Several days later a stack of aluminum siding for use on the 4-H & FFA dormitories caught fire by spontaneous combustion and burned. The $10,000 loss was covered by insurance.[33]

While construction crews raced the clock, Baker's staff hurried to satisfy new demands from exhibitors and vendors. By the end of July advance orders for space soared as sales ran 25 percent ahead of the previous year. Although barn facilities were much larger, fair management ordered eight tents to house the anticipated overflow.[34]

As the construction deadline neared, an editorial in the *Oklahoma City Times* captured the spirit of anticipation:

PUBLISHING
COMPANY
THE DAILY OKLAHOMAN
OKLAHOMA
TIMES
WKY

Despite record rainfall throughout the inaugural fair in 1954, large crowds attended on "Rural School Children's Day" (Courtesy Daily Oklahoman).

STATE FAIR WINNING RACE AGAINST TIME

Despite various handicaps including the prolonged drouth which has held back landscaping, the Oklahoma State Fair has made rapid progress in shaping up the facilities for the annual exposition. A huge assemblage of spick-and-span new, modernistic buildings meets the eye of the visitor who may tour the grounds on the spacious, well surfaced roadway system.

The entire amount available under the bond issue will have been expended by the time the fair opens, and that means all the facilities included in the budget will be ready for use. There are 16 major buildings belonging to the fairgrounds plus a number of attractive structures built and paid for by lessees for their own purposes.

It is hoped that additions can be made in the near future, including a considerable enlargement of the already huge agriculture building plus various recreational and sports features. The first phase—the first go around—is completed according to plan.

It is safe to say that the Oklahoma State Fair is now the most up-to-date and attractive one of its kind in the nation with due consideration to the amplification of some of the structures which is sure to come. If the whole institution can be amplified into a worlds fair according to the dream of many of the supporters, no one will need to apologize for the excellence of the whole group, as to the architecture and utility.[35]

The 1954 State Fair, the first held on the new fairgrounds, was destined to be the largest fair ever held in

the state of Oklahoma to that time. On September 25,
1954—opening day—84,124 people poured through the
turnstiles, setting a new record for one-day attendance.
For the nine-day run, 416,000 people came to see the
new grounds and a greatly expanded state fair. What
made these numbers even more impressive was the fact
that the fair had suffered almost continual rain and a sea
of mud.[36]

Success at the gate did not extend to the financial
sheet. Although the official report cited a gain of
$52,141, the fair was virtually bankrupt. The biggest
blow had been the capital investment of $313,000 in 1953
and 1954 to complete the new site and make the move.
With no cash reserve, there was no cushion to carry on
operations. When bills mounted after the fair, the board
authorized borrowing $75,000 in three increments of
$25,000 each. The loans were made by the First National
Bank & Trust Co. of Oklahoma City and guaranteed by
members of the board of directors.[37]

In 1955 and 1956 the fair continued to lose money,
and it was necessary to increase borrowing to $125,000.
Despite the growing deficits, Pete Baker spent much of
his time in 1955 unsuccessfully trying to develop a baseball trust. Then there was the Semicentennial Exposition, the 50th anniversary of Statehood, scheduled for
the summer of 1957. With much of Baker's energies
focused on expanded or potential programs, the financial
condition of the fair only worsened.[38]

As the crisis mounted, new problems soured relations
with the city, especially a conflict over scheduling of

*The FFA Boys Dormitory was one of
the buildings completed by the time the
first fair opened on the west side
location in 1954 (Courtesy Daily
Oklahoman).*

major events. The city operated the Municipal Auditorium, a multiple-use structure that could be used for a variety of functions from basketball tournaments and boat shows to symphonic orchestras and ballet. When the fair opened with its own multi-purpose facilities, a rivalry for customers naturally arose.[39]

Just as troubling was a feud with Mayor Allan Street, who had proposed building a new Boy Scout Headquarters on the west-side fairgrounds as a memorial to his late son, Bob Street. The Street family pledged a substantial amount to fund the building, but Pete Baker opposed its location on the fairgrounds, stating it would be against the policy of the board. The mayor, incensed, nevertheless guided a proposed lease to the Boy Scouts through the city council. Eventually, the fair board accepted the lease; it had no choice.

On January 24, 1954, there was a joint meeting of the mayor and council with members of the fair board. Mayor Street stated that he had always been interested in the state fair, but felt the city should supervise the fairgrounds since these were municipally owned. He also believed both the council and the state fair needed better public relations, which, in his opinion, could be accomplished by the city having control of the grounds. City manager Bill Gill echoed the mayor, citing several reasons for a new contract. At one point during the emotional meeting, he blurted that if the fair board was not willing to make changes, "There is no use in talking about it." The meeting ended with the fair board authorizing Baker to discuss details before another joint meeting.[40]

The fair board negotiated from weakness. There remained countless projects on the grounds that needed to be completed, but there was no money in the treasury and no prospect for a solution. Then there was the financial obligation of utilities and maintenance on a daily basis. If revenues from the fair would not cover expenses for the actual fair, there was nothing for janitors, utilities, and maintenance the rest of the year.

On January 16, 1957, the fair board granted interim control of the fairgrounds to the city. Thereafter, the fair board would control the fairgrounds one month a year; the other eleven months, the grounds, the buildings, and the maintenance staff became the responsibility of the city. The traditional formula for a public-private partnership, so successful since 1917, was suddenly brought to an end.[41]

The use of the fairgrounds for other groups broadened after acceptance of the lease with the Boy Scouts and the transfer of interim operations to the city. First came the Oklahoma City Round-Up Club, then the Okla-

homa City Science and Arts Foundation and the Oklahoma City Arts Center. All of these new tenants strengthened the mayor's position that the fairgrounds should be run by the city government with the State Fair of Oklahoma only one of many tenants.

The big move, once touted as a promising new start, had turned into a quagmire of mounting debts and political weakness. It would be another 23 years before the fair board regained control and development of the fairgrounds.

The new fairgrounds in 1960, three miles west of downtown Oklahoma City, seen in the background (State Fair Archives).

Chapter Six
COMPLETING THE DREAM

Losing control over the fairgrounds eleven months out of the year confirmed what some fair board members already knew—a change was needed—and needed quickly—if the State Fair of Oklahoma was ever to reach its full potential.

The most obvious problem was facilities. Everyone involved in the planning process, from the Chamber of Commerce to the fair board, realized that the bond issue of 1950 had provided enough capital to build only a fraction of the facilities needed to make the State Fair of Oklahoma a first-class institution. Expansion, they knew, was the key to the future. They also knew that any expansion would have to be undertaken without city financing or interim user fees. Each ten-day fair would have to generate enough revenue to fund every capital project, one brick, one building at a time.

Another problem, one that limited the potential success of any expansion program, was the style of leadership at the fair. Pete Baker was an old-time showman and an all-around nice guy, a man with a heart of gold who would do anything for a friend. He had been with the fair for more than 30 years, and he had been trained by Ralph Hemphill, whose management style and vision for the fair dated to 1918.

Opposite page: Appaloosa class competition at the 1962 State Fair of Oklahoma (Courtesy Daily Oklahoman).

In large part, those very assets of tradition and amiability were at the root of the problem. Postwar America was changing rapidly, both socially and economically, and the changes were especially dramatic in Oklahoma. When Baker had started with the fair, the state was predominantly rural, provincial, and starved for entertainment and a chance to see anything and everything new and modern. By 1960 more people lived in cities than in rural areas, and everyone, including the baby boom generation, had access to things new and modern via television and the movies. Confronted by challenges as well as opportunities, Baker was armed with tried and true traditions that simply did not work any longer.

In normal times Baker probably could have adjusted with direction from the fair board, but the 1950s were not normal times. First, he had been asked to organize and execute a once-in-a-century move, abandoning an established fairgrounds for a rough, undeveloped parcel of land where bogs had to be drained and basic utilities had to be installed. And despite the general consensus that the new fairgrounds had been undercapitalized, both the fair board and the public had high expectations

Ista, the "Butterfly Dancer," was the feature act at Club Lido on the midway in 1960 (Courtesy Daily Oklahoman*).*

In 1958, with the world focused on space exploration, the State Fair of Oklahoma featured a personal appearance by the actors who played the television personality of Buck Rogers (Courtesy Daily Oklahoman).

A young exhibitor from Beaver at the 1970 livestock show (Courtesy Daily Oklahoman).

after the move. It was neither the time nor the place for a nice guy comfortable with the past.

The task of charting a new direction fell to an unusually capable fair board. Included in the ranks were veteran board members such as C.R. Anthony, founder and chief operating officer of a regional clothing chain bearing his name; Virgil Browne, founder and president of Oklahoma Coca Cola Bottling Company; Luther Dulaney, founder and president of the state's largest distributor of appliances and electronics; Donald Kennedy, president and chairman of the board of Oklahoma Gas and Electric Company; John Kilpatrick, Jr., owner of a lumber company and extensive real estate holdings; John Kirkpatrick, president of Kirkpatrick Oil; Lou A. Macklanburg, founder of Macklanburg-Duncan Industries; and Dean A. McGee, president of Kerr-McGee Oil Company.

It was a distinguished group, armed with a wide range of skills and executive experience in organizing resources and directing people. The only weakness was the size of

the board. Meeting only quarterly, the full board simply was too far removed from daily operations of the fair to identify problems and articulate solutions, especially when the top manager was part of the problem. The battle flag of leadership, instead, had to be carried by the executive committee.

By 1959 the executive committee had become an increasingly assertive part of the fair's management team. This was due in large part to the growing complexity of the operations, but it also was a result of Baker's weak management. Fortunately for the long-term development of the fair, the executive committee in 1959 included members well prepared for the challenge.

The president was Dan James, owner of three hotels—the Skirvin, the Skirvin Tower, and the Black—where service was paramount and the common goal was a high standard of quality. Other members included Nelson Owen, vice president of Oklahoma City Federal Savings and Loan; Larry Wolf, who owned an advertising agency; and Stanley Draper, managing director of the Oklahoma City Chamber of Commerce and one of the most respected urban planners in the nation.   The youthful, dynamic personality of the executive committee, however, was best represented by its vice-president, a businessman who would soon become president of the fair board and the architect of its dramatic emergence as a world class institution. That man was Edward L. Gaylord.

In 1959 E.L. Gaylord was executive vice-president and assistant general manager of the Oklahoma Publishing Company, the parent company of a communications empire that included the *Daily Oklahoman*, the *Oklahoma City Times*, the *Farmer-Stockman*, WKY-TV, and WKY-Radio, as well as a number of affiliated companies that ranged from trucking lines to paper mills. As a young man on the rise, he had already become president of the Chamber of Commerce and Oklahoma Industries, Inc., and he had served as director on a number of local civic boards such as the Oklahoma Zoologicical Society, the Oklahoma Medical Research Foundation, Oklahoma City University, the YMCA, and the United Fund.

In addition to the advantages inherent in his civic and business connections, Gaylord brought to the fair board a personal background in the three most important parts of the fair's mission—boosterism, education, and enter-

Tinker Air Force Base, established in 1941 as Oklahoma's largest employer, sponsored displays at the State Fair of Oklahoma to explain its role in the national defense system. This elaborate entry gate was erected at the 1963 fair (Courtesy Daily Oklahoman*).*

tainment. He was an advocate for free enterprise and economic development, a conviction inherited from his family and nurtured through formal training that included a bachelor's degree from Stanford and an MBA from Harvard Business School. Surprisingly to some people, this interest extended to agriculture, the mainstay of the fair's programs. As a young man, Gaylord had spent considerable time on the family farm, where his father raised dairy cattle. Until the age of fourteen, young Gaylord was intent on being a farmer; his father persuaded him to reconsider that choice of careers.

He also was a strong supporter of education in its many forms. He was an advocate for Oklahoma City Public Schools, which he had attended, as well as for Oklahoma City University and Oklahoma Christian College. Later in life he would extend this financial support through large donations to both The University of Oklahoma and Oklahoma State University. Interested in innovative approaches to education, he helped found two heritage museums, the Cowboy Hall of Fame and Enterprise Square.

Maintaining the role of community booster, the State Fair in 1966 displayed the scale model of the Pei Plan, a long-range concept for the revitalization of downtown Oklahoma City (Courtesy Daily Oklahoman).

In the field of entertainment, Gaylord's experience with radio and television ultimatately led him into movie production, a cable television network, ownership of Opryland, USA, part ownership of the Texas Rangers, and the creation of Fiesta Texas, a theme park that combined entertainment and heritage education. Through these varied enterprises, he developed a keen sense for what people wanted to see, hear, and experience.[1]

Armed with this wide range of personal and professional experience, Gaylord willingly applied his talents to moving the State Fair of Oklahoma to the next level of development.

Initially, Gaylord and the executive committee moved to support rather than replace Baker. Seeking a "closer working relationship between the Board of Directors and management," they created a new committee structure. Heading the list was the Finance and Facilities Committee, assigned the task of an overall plan for the operational budget as well as a master plan for expansion. Others included the Livestock Committee, an Agricultural Committee, a Youth Activities Committee, a Business and Industry Committee, an Arts and Science Committee, and an Attractions and Special Events Committee. The new planning structure not only provided more oversight of daily operations, but also defined once again the primary objectives of the fair itself.[2]

Committee support, unfortunately, fell short of accomplishing the needed changes. When the 1960 fair lost more than $70,000, Gaylord was assigned the task of meeting with Baker to ask him to step down. In a letter dated January 17, 1961, Baker dutifully asked that the executive committee "secure the services of a dynamic younger man with broad experience in fair management to succeed me as Secretary-Manager at the earliest practical date." The fair board agreed to pay him full salary for another year to serve as a consultant to "guarantee a smooth transition." Thereafter, he would be paid half his salary as a retirement benefit.[3]

Gaylord, by this time chairman of the executive committee as well as president of the fair board, led a national search for a new secretary-manager. He wanted a man familiar with the state and its people, a man who understood agriculture, and, most importantly, a man who would bring an aggressive approach to fulfilling the basic mission of the fair while generating enough income to build for the future. On March 20, 1961, Gaylord announced to the fair board that the committee had found such a man—Orval O. "Sandy" Saunders.[4]

Saunders was born on December 13, 1916, in Spokane, Washington, the son of a lumber mill operator. After moving as a youth to Montana, he worked for the Forest

Rodeo producer Jiggs Beutler (standing with cowboy hat) visited with Russian scientists at the 1961 fair (Courtesy Daily Oklahoman*).*

Oklahoma astronaut Gordon Cooper's space capsule was the object of attention for his mother, Mrs. Hattie Cooper (right), his grandmother, Mrs. Orena Herd, and a NASA official in 1963 (Courtesy Daily Oklahoman).

Service, then joined the army during World War II and rose to the rank of lieutenant colonel. After the war he moved to Oklahoma and found a job with WKY Radio as assistant farm director; within three years he was director. When WKY-TV went on the air in 1949 he served as farm director for both stations, doing a 30-minute live broadcast on television as well as two radio shows every day.[5]

In the mid-1950s Saunders left WKY to become state manager of the American Dairy Association, and soon was promoted to its national headquarters in Chicago. During these years he added new skills to his arsenal of experience. He learned how to influence public opinion through trade exhibits and came to appreciate the close relationship between educational outreach and economic development. All of this was added to his knowledge of Oklahoma's farm economy, gained while he was farm director at WKY. When Gaylord and the executive committee interviewed Saunders for the job at the fair, they recognized that all of this experience would blend well with his aggressive, "get the job done" attitude.[6]

When Saunders arrived for his first day on the job, he quickly recognized the physical and financial challenge. Streets were unpaved, landscaping was nonexistent, and buildings central to the success of the master plan had been put on indefinite hold. Then there was the deficit of $70,027, an operational shortfall that crippled even

the most modest attempts at making desperately needed
capital improvements. With no bond money left, an
indifferent city council, and a cloud of debt hanging over
the horizon, the fair board had to develop a new strategy
for moving ahead.[7]

First, the executive committee committed itself to
tightening financial control. As Saunders later recalled,
"there was a gravy train at the fair...freebie stuff and
discounts." They "unloaded" the gravy train and adopted
concession and exhibit rates that were consistent and fair.
When they examined the books for the previous two
fairs, they discovered that the midway concession company had been granted an extra cut of the revenues,
simply because it had convinced the soft-hearted Baker
that it needed the money to survive.

Saunders, citing the original contract, explained that
the discount had not been approved by the board, and
therefore an additional $6,180 had to be paid to the fair.
He was backed by the executive committee, and the bill
was paid. With similar aggressiveness, the fair board
rewrote insurance policies, put the printing of tickets out
for bid at a savings of 40 percent, and raised fees for
independent game and carnival operators for the next
fair. If the fair was to grow and prosper, the board knew
that every nickel and dime had to be carefully nurtured.[8]

With better financial control, the small staff of Saunders, Frances Young, Marie Sullivan, Lem Jones, Cliff
Norgard, and Earl Schweikherd turned its attention to
improving exhibits and attractions, the surest means of
generating bigger crowds and higher earnings. First came
simple adjustments, such as a new emphasis on cleanli-
ness. For starters, they placed an attendant in every rest-
room just to keep facilities clean. More challenging was the
problem with exhibitors, the fair's traditional attraction.

*By 1961 the physical plant of the
fairgrounds included two additional
buildings funded by John and Eleanor
Kirkpatrick, the Science and Arts
Museum (on the right) and the Art
Center and Planetarium (upper right)
(State Fair Archives).*

Farmers and ranchers, the executive committee learned, were unhappy with the fair. They had not been getting expected amenities and services, and as a result the number of entries had dropped along with general attendance. Immediately, with the fair board's financial backing, the staff initiated construction and renovation of stalls, gave added force to demands for a show arena, and provided simple things like plentiful straw and water. "We listened," said Saunders, "and it worked."[9]

Gaylord and Saunders then turned their attention to commercial exhibitors, an even better source of revenue that also attracted crowds. They hired a salesman who did nothing but sell to area businessmen. "It was a trick to get to the point that a salesman could sit down with the manager of a private company and convince him that he should show his products at the fair," said Saunders. They printed four-color brochures and, more importantly, got to potential exhibitors early. Encouraged by initial success, they added two more salesmen. "The

Lomega FFA Chapter members Diane Schwarz and Robin Bernhardt led their sheep to the barn in 1977 (Courtesy Daily Oklahoman).

whole program worked," he said afterward, "and the fair began to grow."[10]

In one of his first newspaper interviews after arriving in Oklahoma City, Saunders gave an indication of the fair board's willingness to build. "The first impression I got," he said, "was a simple lack of facilities to put on the kind of show Oklahoma would like to have." Indeed, the fairgrounds, with more than 400 acres of land, did seem underdeveloped. There was the Grandstand on the east side, livestock barns on the far west side, the 4-H and FFA buildings on the north end, and only the General Exhibits Building, the Appliances Building, and the Oklahoma Publishing Company Building in between. As Saunders later recalled, "I could look from my office in the grandstand all the way to the livestock building without anything blocking my view."[11]

The only construction on the grounds since the original move in 1954 had been financed by John and Eleanor Kirkpatrick, well known philanthropists who were just beginning a long career of investment in the community. On December 5, 1958, the fair had dedicated the Kirkpatrick Art Center, a $250,000 exhibition building and planetarium billed as a "versatile teaching tool, usable from the third grade upward...a cultural tool of the first importance." Then came the Kirkpatrick Science and Arts Museum, a 42,000-square-foot building constructed at a cost of $400,000 in 1962. It included space for workshops, labs, art classes, ballet studio, auditorium, enlarged planetarium, as well as offices for the Ladies Music Club, Frontiers of Science, the Oklahoma City Charity Horse Show, and the Science and Arts Foundation. Although the two buildings were privately funded with ongoing programs administered by independent boards of directors, both were considered part of the fairgrounds operation during each fair, with exhibits chosen by the fair board.[12]

A crew from the 2953rd Combat Logistics Support Squadron towed a C47 under the I-40 overpass at MacArthur Avenue on the way to the State Fairgrounds. The craft, which had flown 900 hours transporting paratroopers during World War II, was donated to the fair by Kerr-McGee Corporation (Courtesy Daily Oklahoman*).*

Initially, the executive committee was confronted by the proverbial dilemma of which comes first, the chicken or the egg. Without additional exhibit space, they could not generate large amounts of cash, and without large amounts of cash they could not add needed exhibit space. While belt tightening and better management held out hope for the long run, there had to be an immediate effort to build attendance and revenue at the gate. The most cost-effective solution, they discovered, was a heavy reliance on outdoor exhibits, especially military hardware and space age technology.

As many members of the fair board knew, Oklahomans had long been interested in weaponry. During

both World Wars I and II military hardware, from diri-
gibles to the first armored tanks, had been popular attractions at the fair. Mock battles were staged, and the
latest in submarines, aircraft, and artillery were on display for those left on the home front. Then came the
Cold War, the Korean Conflict, the Berlin wall, the
Soviet invasion of Hungary, and the spread of Commu-
nism in Southeast Asia. Worst of all was the mounting
threat of nuclear holocaust. In 1957, when the Soviet
Union launched Sputnik, Americans were suddenly
confronted with the realization that the oceans no longer
guaranteed isolation from the armed conflicts of the
world. For the first time since the Civil War, American
cities were vulnerable to foreign attack.

The ability to use outer space for the delivery of
increasingly destructive weapons shocked Americans, so
the federal government rushed to build new weapons
systems, both for ground and air war, and developed
promotional and educational exhibits to fan public support. The result was a growing public fascination with all
things military and "high tech." The State Fair of Okla-

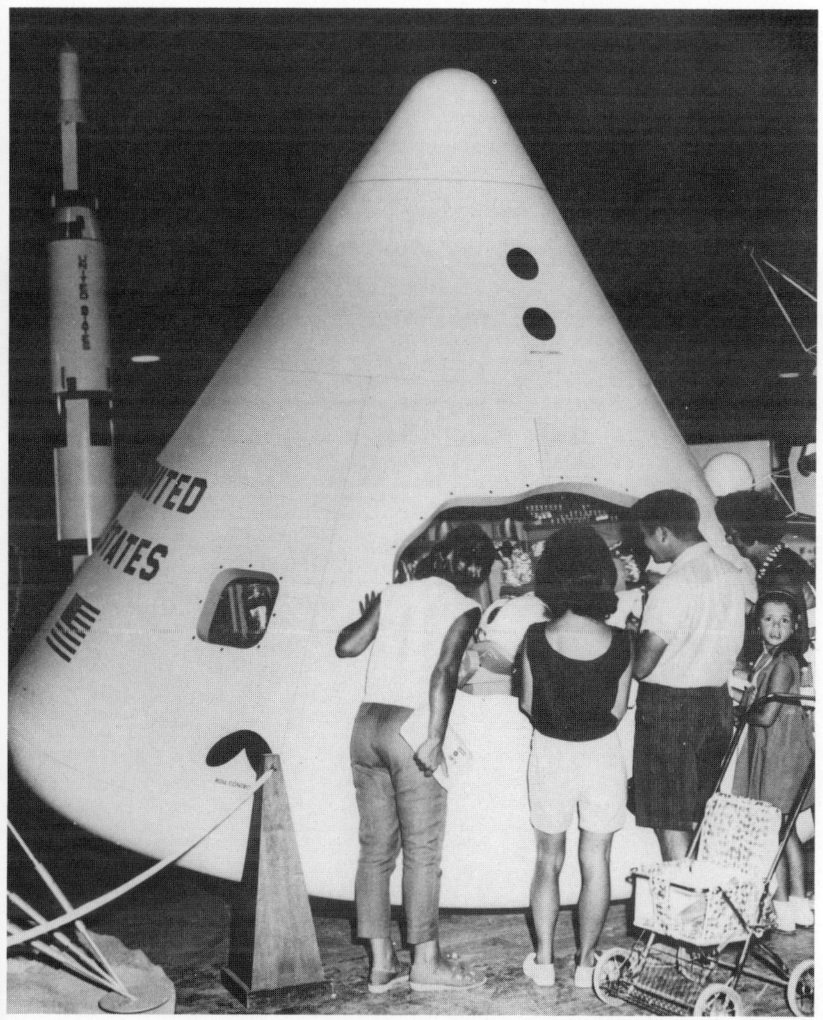

This official replica of an Apollo command module was the featured attraction at NASA's "Space Flight Center" exhibit at the 1967 fair (Courtesy Daily Oklahoman*).*

homa, needing cheap outdoor exhibits to attract greater crowds, was the natural agent to satisfy that curiosity.

Advertisements for the 1961 fair promised a "Big Space Show." The armed services were there in full force, circling the central plaza with a nose section of a B-52, an M-40 tank, and a display that included Scout, Sergeant, Corporal, LaCrosse, Honest John, Little John, Nike-Hercules, and Nike Zeus missiles. A big attraction was a firing demonstration of a Nike-Hercules with crews operating at both slow motion and combat speed.[13]

The Civil Aeronautics Administration, which had moved its training facilities to Oklahoma City in 1948, also was present at the "Big Space Show." Exhibits included scale models of airfield towers, electronic landing equipment, navigational aids, and radar. Visitors to the display could even listen to live communications between pilots and air traffic controllers at Will Rogers World Airport.[14]

Their sights set even higher, Saunders and fair board

Dean McGee, president of the fair board from 1971 to 1977, views a scale model of a missile base on display at the fair (State Fair Archives).

members used their contacts in Senator Robert S. Kerr's office to meet with public affairs staff at the National Air and Space Administration. Previously, they had limited their educational and promotional displays to universities and museums, but sensing the opportunities of reaching even greater numbers of people, NASA agreed to send something to the fair in Oklahoma City.[15]

The NASA exhibit included scale models of the Echo communications satellite and the Tiros weather satellite. The centerpiece was a multicolored globe eight feet in diameter showing the orbits of all satellites circling the earth. Outside was a ten-story-tall Scout rocket used to boost satellites into orbit. When coupled with the other exhibits, especially the popular "Venture into Space" show at the Kirkpatrick Planetarium, the fair provided a rare insight into the changing world of space exploration and high technology.[16]

Taking full advantage of this public interest, the fair board chose "Galaxy of Progress" as the theme for the 1963 fair. One exhibit featured Gordon Cooper's Faith 7 space capsule, while Martin Marietta Corporation, one of the largest defense and industrial firms in the world, sponsored a "Moon Base Exhibit" that simulated conditions of life in space and the likelihood of a moon base colony.

The race to the moon remained a lively topic throughout the decade. In 1970 the Kirkpatrick Planetarium offered a 45-minute show called "The Man, the Moon, and Green Cheese," with slides of the lunar surface, rockets in flight, and the origin of the planets and stars. The next year, NASA provided an even more lifelike setting with its "Put Yourself on the Moon" exhibit. A sandy moonscape diorama dotted with molten glass, rocks, and volcanic debris was the setting for a moon rover and two actors dressed like Neil Armstrong and Buzz Aldrin.[17]

NASA, needing good public relations to insure healthy appropriations from Congress, provided a seemingly endless supply of educational exhibits. In 1964 a 4,000 square-foot display titled "Gateway to Space" included models of a Gemini space capsule that at that time was being used to orbit the earth and an Apollo space capsule that eventually would take man to the moon. Also included in the exhibit were full-scale mock-ups of Nimbus, the latest weather satellite, Explorer II and Pioneer V, and Syncom II, billed as a modern communication satellite that would "shortly bring us the Olympics from Tokyo."[18]

Another source of cost effective, popular exhibits was Tinker Air Force Base. Tinker traced its history in the community to 1942 when the Army Air Corps had estab-

An eight-foot clock was installed 100 feet above the ground at the top of a galvanized steel oil derrick in 1977 (Courtesy Daily Oklahoman).

lished a string of air depots across the country. The base survived World War II and even expanded as the growing Cold War kept the country at a high level of preparedness. Tinker, the largest employer in the state, had both the resources and the incentive to teach the public more about the military and its role in American society.

As early as 1960 Tinker sponsored an "Aero-rama," with scale and working models of missiles, aircraft, radar, and a working weather station. Base employees also showed a series of films on Tinker's operations. The next year Tinker sponsored the display of a GAM-77 Hound Dog missile described in newspaper accounts as "a deterrent missile carrying a nuclear warhead that will be used in case of enemy attack on this country." Gaylord, Saunders, and the fair board readily recognized the potential of both displays.[19]

Over the next two decades Tinker sponsored a seemingly endless number of new aircraft. One of the most popular was the supersonic X-15, the "first winged space plane," placed on display in 1962. Others included the T-38 Talon, a jet that could fly more than 800 miles per hour and climb 30,000 feet per minute to an altitude of 55,000 feet. Then came the F111A, complete with a mock cockpit where visitors could sit and work controls. An even more realistic experience was a nine-minute 3-D film that put the viewer in the pilot seat of a jet plane taking off from an aircraft carrier to make a shore bombardment.[20]

By 1970 officials at Tinker agreed to place permanent exhibits in the plaza just north of the Transportation Building, an area that had been assigned to Tinker for more than a decade. Their intent was to pull together a collection of aircraft that would represent the development of Tinker, specifically a B-47, a B-52, and a DC-3. The first acquisition, a B-47, weighed 50,000 pounds, measured 109 feet from nose to tail, and had a wingspan of 117 feet. Placed on concrete piers, it opened for public view in 1971.[21]

A more difficult challenge was the B-52. The giant bomber was flown into Tinker where crews took off the wings and tail. The only way to get it to the fairgrounds, however, was to roll it on its own wheels behind a truck. It was towed west on 29th Street, north on Western to Reno, then west onto the fairgrounds. Along the route, traffic lights and power lines had to be removed, especially at corners where the long-range strategic bomber had to be turned. By the time it was installed on its piers, the cost of the project exceeded $55,000.[22]

In 1976 the fair board constructed the log stockade to provide free entertainment in an open-air theater (Courtesy Daily Oklahoman).

Above and opposite page: In the 1960s, with the Arena and a better sound system, the State Fair Rodeo turned to musical entertainment with stars such as Roy and Dale Rogers, seen here in 1968 (State Fair Archives).

A renewed emphasis on cost-effective exhibits combined with a more aggressive, efficient management style paid handsome dividends. In his first year as president of the fair board, Gaylord and the staff erased the $75,000 deficit from the 1960 fair and generated a surplus of $83,000. Over the next five years, attendance climbed from 455,671 to 831,525, while net income from exhibitors, concessions, and entertainment soared even more quickly, from $108,724 to $304,149. Armed with a healthy cash flow, a supportive board, and a motivated, committed staff, the State Fair of Oklahoma embarked on the greatest period of capital expansion in the fair's history.[23]

"Completing the dream" began in October of 1961, when the fair board announced an eight-year, $4.6 million capital expansion plan. The first phase, to be completed by 1964, included a Transportation Building, estimated to cost $1.2 million, a $225,000 addition to the Women's Building, an addition to the Agricultural Building, more horse barns, and, most ambitiously, a multi-purpose arena estimated to cost $2.8 million.[24]

To most board members, the arena was the top priority, for it would benefit all aspects of the fair's program. "The horse industry in Oklahoma is big business," stated the updated master plan, "and the inclusion of an arena provides not only the opportunity for performance classes of horses, but a facility for World Championship rodeos, judging finals in both stock shows and the fair, national livestock breed shows, and the Sooner Dairy Show." In addition, one board member predicted, an arena would offer the chance to attract "circuses, ice shows, ice hockey, basketball, stage productions, and large conventions." The sales pitch worked. By the following October, the arena was included in a city-wide bond issue that also provided funding for the construction of a downtown convention center and a renovated Municipal Auditorium. To help promote the bond issue, the fair board appropriated $5,000 for publicity.[25]

While they waited for the results of the bond issue election, the fair board pushed ahead with other capital projects. They borrowed $145,000 from the First National Bank to build two barns, one a 170 x 360 foot structure to house a Children's Barnyard, poultry exhibits, and 350 head of cattle. The other barn, measuring 100 x 70 feet, was designed for sheep. The added exhibit space allowed more entries and solved a scheduling problem that had previously split the competition into two phases, the first for youth, the second for adults. For the first time since the move from the old fairgrounds, both levels of competition were run concurrently in 1962.[26]

The pace of construction continued into 1963 and 1964 funded by revenues generated by the fairs in 1961, 1962, and 1963. The largest addition was a 62,400-square-foot horse barn, with judging ring, bleachers, 220 stalls, and paved public areas. Other notable projects included the paving of streets and sidewalks, a 14,000-square-foot addition to the Home Arts and Crafts Building (formerly called the Women's building), a 186 x 220 foot addition to the Dairy Barn, and a 14 Flags Plaza around the Arrows to Atoms tower complete with 40-foot flagpoles, seats, sidewalks, and extensive landscaping improvements. President Lyndon B. Johnson attended dedication ceremonies for the flag plaza, an event that drew 30,000 people on September 25, 1964.[27]

The most dramatic leap forward came in 1963 when the citizens of Oklahoma City approved a bond issue that included money for the arena. As designed by the architectural firm of Jack L. Scott and Associates, the oblong-shaped arena was to be 70 feet high, 402 feet long, and

Bucking stock and flying cowboys at the State Fair Rodeo (Courtesy Daily Oklahoman*).*

318 feet across. The exterior wall would be formed from 42 concrete columns which would support a roof suspended on a network of cables ten feet apart. As a result, the roof would be lower at the center than on the sides.[28]

As work progressed on the $1.8 million structure, it quickly became apparent that the bond money would not cover the cost of interior details and special features. The city's share of the funding was increased to include a tunnel linking the arena to the horse barns, but this did not provide for chutes, gates, or fences needed for rodeos and horse shows. The fair board paid the difference. Other add-ons paid by the fair board included $25,000 for a sound system, $13,500 for dressing rooms, $27,000 for a removable basketball floor, $109,030 for box seats and folding chairs, $34,000 for a scoreboard, $110,000 for portable ice-making equipment, and $160,000 for paving the parking lot. To raise the cash,

Before the Arena was completed, the State Fair Rodeo was staged in the infield of the Grandstand. At the 1956 fair, cowboys competed for a total purse of $15,000 (Courtesy Daily Oklahoman*).*

the fair board borrowed $223,000 and used revenue from the past fair for the rest. By opening day of the 1965 fair, the Arena was ready for use.[29]

The Arena may have had the biggest single impact on the fair, but it was only one of many achievements made in the early 1960s. Every year thereafter, without exception, the fair reported progress on some aspect of development, whether it was paving another street or installing more lights. One ongoing investment that would remain a high priority for years was air conditioning. Prior to 1965 none of the major exhibit buildings had been air conditioned, an inconvenience both the public and exhibitors had always accepted. But times were changing, and most people expected air conditioned comfort, even in the fall when Oklahoma temperatures usually dropped a few degrees.

In 1966 the fair board voted to install air conditioning units in all buildings, a major decision given not only the original investment but also the monthly utility bills. Air conditioning in the Arts and Crafts Building in 1967 alone cost $61,000. By 1970 all of the older buildings, except the livestock barns, had been climate controlled.[30]

Another ongoing investment during the 1960s and early 1970s was landscaping. For the first few years after

The fair board purchased the Frizzell Coach Works for display in the Transportation Building (Courtesy Daily Oklahoman).

he became president of the fair board, Gaylord and the staff had to spend most grounds improvement monies on basics such as sidewalks and street paving. In 1964, for example, the fair board paid for one permanent street 40 by 600 feet through the livestock area, a service street 20 by 300 feet between the livestock pavilion and the beef barn, 12,880 square feet of walkways in the beef barn, and 6,992 square feet of walkways in the livestock pavilion. Little was left from the annual budget for extras such as trees, shrubs, and flowers.[31]

A flurry of landscaping activity began in 1966 when the fair board voted to fund the 50 States Flag Plaza. The investment not only improved the visual appeal of an important area between the 4-H and FFA buildings, but also drew new attention to grounds beautification. In March of 1967 the fair board and Oklahoma City Parks

Department planted 178 trees on the grounds; two years later another 290 trees were planted. Other grounds beautification efforts over the next few years included 386 maple trees planted in 1972, and the largest single improvement, the planting of 1,776 redbud trees during the Bicentennial Celebration.[32]

One of the largest single investments in the visual appearance of the fairgrounds was the lighting and luminary project begun in 1969. The fair board funded the purchase of more than 50 multicolored outdoor lights surplused after the New York World's Fair. City crews did the installation. Including safety lighting around all buildings and parking lots, the investment totaled more than $154,000.[33]

In 1976, to celebrate the Bicentennial, the fair board invested another $150,000 on outdoor beautification

A partnership of the fair board and the Oklahoma City Parks Department planted 75 new trees around the Plaza of the States in 1967 (Courtesy Daily Oklahoman*).*

projects. The centerpiece was Constitutional Fountain and Independence Arch. The fountain, one of five installed that year, was 56-feet in diameter and sprayed more than 2,800 gallons of water per minute. Around the face of the fountain was the entire Preamble of the Constitution of the United States. Soaring overhead was an 80-foot gateway arch emblazoned with the first words of the Declaration of Independence, "When in the Course of Human Events...." An added salute to history that year was the State Fair Stockade, a log fort measuring 120 x 240 feet built at a cost of $96,000. Inside were commercial exhibitors and a stage that featured Indian dancing, square dancing, old-fashioned fiddling contests, and the "World Championship Fast Draw Contest."[34]

While constant efforts were made to improve the general appearance of the grounds, Gaylord, Saunders, and the fair board made a final push to complete all of the buildings included in the master plan. In 1967 work finally began on the Made in Oklahoma Building, with 33,400 square feet of space constructed at a cost of $242,000. Then came the long-awaited Transportation Building, built in two phases, with 80,400 square feet of space. The total cost, including air conditioning and a concrete and brick exterior, was $627,000, the largest single investment ever made by the fair board. Both buildings were quickly filled with exhibits, from automobiles and airline booths to international exhibits from 16 countries.[35]

This volunteer from Carl Albert High School in Midwest City was one of 70 students who kept the Children's Barnyard open at the 1972 fair. Housed in a tent, the boys slept on cots behind a partition (Courtesy Daily Oklahoman).

The decade of the 1970s closed with a steady stream of other improvements to the grounds. For public convenience, several new restrooms were constructed at a cost of $200,000. Two more bombers, a B-52 and a B-47, were installed as permanent exhibits at a cost of $54,000. The clock tower went up in 1977 at a cost of $42,000. And several additions were made to the livestock area, including a horse exercise arena, a new milking parlor, and permanent stalls in the horse barn.

From 1961 to 1980, including all projects, the fair board reinvested $6,910,834 from earnings. All of these investments, whether simple additions such as sidewalks or complex projects such as new buildings, meant that the fair was getting closer to completing the dream envisioned at the time of the move from the east side. It also meant that the fair finally had the physical facilities to host a well-balanced program that everyone, especially Sandy Saunders and Edward L. Gaylord, wanted. It was an opportunity they would not miss.

By 1965 the fairgrounds was filling out, with the baseball stadium (lower right), Arena (upper right), and the monorail (State Fair Archives).

Chapter Seven

ALL THINGS TO ALL PEOPLE

As proven by the financial and physical achievements of the 1960s and 1970s, the secret to success at the State Fair of Oklahoma was the ability to adapt to challenges as well as opportunities.

Financially, the fair had been losing money in the 1950s, so the fair board hired a man who would run the fair like a business, maximizing revenue while holding down expenses. Physically, the fairgrounds had needed major investments after the big move was completed, so the board and staff crafted a long-range plan that balanced steady improvements against revenue. Neither of these challenges could have been met, however, without a basic change in the way the fair was marketed to the general public.

"We had a strong belief," Gaylord later would say, "that the fair had to have something for everyone. It had to be all things to all people, whether they were city folks, teenagers, or homemakers." It was a concept well fitted to the times as well as the immediate financial needs of the fair.

The fair board knew that if exhibits and programs appealed to a broad audience, attendance would increase. And if attendance increased, revenue would go up, allowing the booking of better entertainment and the

Opposite page: Junior competitors in the lightweight Suffolk market lamb contest in 1969 (Courtesy Daily Oklahoman).

Edward L. Gaylord, president of the fair board from 1961 to 1971, led the fair in its greatest period of expansion (Courtesy Daily Oklahoman).

construction of more exhibit space. The fair board, along with Saunders, understood that the ultimate reward would be greater public support and a cash flow to fund improvements to the grounds and facilities.[1]

Initially, they applied this new criteria of "all things to all people" to the traditional features of the fair, such as livestock shows and agricultural displays. In both of those program areas, it was not so much a change as it was a rejuvenation, a badly needed infusion of resources and attention. The results were dramatic. Despite the rapid pace of urbanization in the state, the dairy exhibit in 1979 was the largest in the history of the State Fair with 539 head of milk cows from seven states. This was in addition to an expanded beef cattle competition that included 650 head of Charolais, Angus, Herefords, Polled Herefords, Santa Gertrudis, Brangus, Limousin, Gelbrieh, Maime-Anjou, and two new breeds, the Blonde d'Aquitane from France and the Cheanina from Italy. There also was a children's petting zoo and a 4-H compe-

The Livestock Pavilion was the only structure moved to the new fairgrounds from the east side location (State Fair Archives).

A blending of old and new greeted this 4-H student competitor at the 1970 State Fair of Oklahoma, where livestock shows ran alongside new events such as the Ice Capades (Courtesy Daily Oklahoman*).*

tition that had never been stronger. Clearly, attention to marketing helped keep agricultural displays among the top attractions at the State Fair of Oklahoma.[2]

Whereas the agricultural program of the fair needed a tune-up, the midway needed an overhaul. For decades the midway concession had been given to Royal American Shows, a traditional amusement company that featured freak shows and exotic dancers such as Sally Rand, a well-known stripper who played the state fair as late as 1960. Royal American also had been slow in adjusting to Baby Boomer demands for more rides; the few rides the company did offer were tame by modern standards. Even worse, in the opinion of Gaylord and the fair board, the Royal American midway did not make enough money for the fair.

The relationship with Royal American began to unravel in 1961 when the executive committee forced it to pay its full fee from the 1960 fair, even though Pete

Rod Link provided a new look for the Midway by 1968, with more emphasis on rides and less on sideshows (State Fair Archives).

Baker had given the company a financial break, albeit without board approval. Then in 1967 Royal American did not generate enough income to pay the fair more than the minimum allowed by contract. When carnival managers blamed the shortfall on the fact that three of their featured rides had been out of commission for at least three days for repair, Saunders told them they should have anticipated the problems and done the repairs before the fair opened. When he demanded they pay a percentage of the lost revenue, Royal American balked and announced it no longer would play the State Fair of Oklahoma.[3]

With input from the fair board, Saunders hurriedly put together a bid package that included the dates of the fair, attendance figures, and dollar income, then sent it to all major carnival companies with instructions to submit their bids to him at a national convention in Chicago. He quickly rejected bids that seemed too good to be true. "To pay 60 percent of the gross...they either had to steal from the public or steal from us," was Saunders' rationale. After interviewing all applicants, he chose to work with Rod Link, owner of the Amusement Corporation of America, who offered a $100,000 minimum with 42 percent of the first $300,000 and 45 percent of the gross above $300,000. By the next fair, with no more rides than the previous carnival, Link increased the fair's "part of the action" by more than $120,000.[4]

Link's operation grew and changed according to the needs of the fair. As the Baby Boom generation entered their teen years and the midway became the public's number one destination, the demand for thrill rides

escalated, so Link added rides such as the Sky Diver, the
Zipper, the Zyklin, the Speelunker, and the Swiss Bob,
rides that went faster and flew higher. By the early
1970s he was offering 120 rides, requiring a fleet of 190
trucks just to get them from one location to the next.
More importantly, the "gross" increased to $1.6 million.[5]

Finding a new midway company was only one of the
many ways the fair was changed by the Baby Boomers,
those kids born after 1946 who were entering their teen
years by the early 1960s. To lure this affluent, free-
spending generation of teenagers, the fair board created
specific programs that appealed to them.

In 1966 the fair board arranged for a "Teen Fair"
with a "teen board" of students selected from city schools
to organize the program. That first year, a tent was set
up with a stage for musical groups and physical fitness
demonstrations. Typical of the new approach to programs that paid for themselves, the plan included exhibit
space for vendors selling youth-oriented items such as
guitars and "funny hats."[6]

The Teen Fair expanded the next year into a 10,000-
square-foot building, called the "groovy barn," with more
exhibit space for vendors pushing youth-oriented goods

*The Starship Enterprise at the 1981
fair (Courtesy* Daily Oklahoman).

Responding to the Baby Boomer generation, the fair board encouraged Link and independent contractors to bring bigger and faster rides to the State Fair of Oklahoma (Courtesy Daily Oklahoman).

Below and opposite: Daily parades added color and pageantry to the celebration of the American Bicentennial (State Fair Archives).

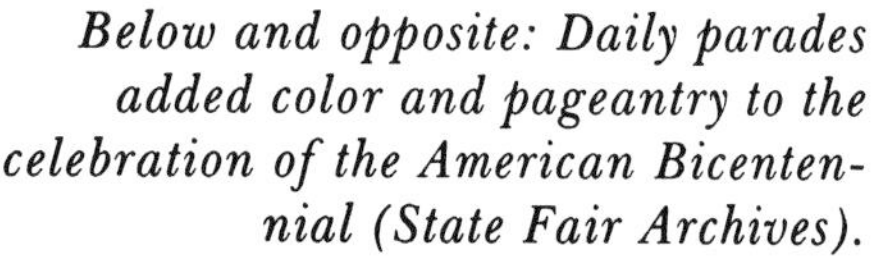

such as musical instruments, records, motorcycles, and cosmetics, as well as schools, the Job Corps, movie theaters, employment offices, and churches. A "psychedelic paint-in" was added in 1968, when an old house was brought in so students could, in the words of one reporter, "express themselves in their own manner." TG&Y, a local department store, provided the paint and brushes.

The 1968 teen fair also featured a "psychedelic show." In one house, called "Something More," teenagers walked through rooms filled with eucalyptus smoke, colored strobes, black lights, "hippie posters," and kaleidoscopic projections. Next door was a house, called "Something Else," with a ten-minute light show set to rock music that had won first place at a psychedelic film festival in Bonn, Germany.[7]

Another impact of the Baby Boom generation, but one that was more a reaction than an accommodation, was a growing emphasis on patriotism in the fair's programs. As the war in Vietnam rose to a controversial climax in the late 1960s and early 1970s, the fair board attempted to heal the wounds of dissension and reaction. In 1969 the theme of the fair was "Adventures in Patriotism," and included a grand ceremony each day with honor guards from 32 military and civilian organizations. As they stood at attention, World War II veterans raised a 20-by-30-foot flag while a band played the National Anthem.[8]

An even greater expression of national pride was the Bicentennial Celebration featured at the 1975 fair. Called "The Spirit of America Exhibition," the daily ceremonies started with the raising of a 27 x 42 foot Old Glory and the firing of a 50-gun salute. Each afternoon at 5:30, a parade was staged with a Marine Corps Color Guard, an Old Guard Fife and Drum Corps, costumed riders, four 100-year old carriages, and a troop of Revolutionary War soldiers. The grand celebration of the nation's founding also included a "Constitutional Fountain" with the Constitution engraved around the perimeter, 1,776 redbud trees planted by Boy Scouts, and an official visit by President Gerald Ford. "It's patriotism we're selling," said a proud Sandy Saunders as he watched yet another new attendance record set.[9]

The policy of "all things for all people" was clearly evident in the concurrent effort to cater to both free-spirited teenagers and more traditional adults. Another result of this universal appeal to all visitors was a labyrinth of independent carnival games and food vendors.

Games had been part of the fair for years, starting with shooting galleries and simple throwing contests such as knocking down "Cats" or stacked milk bottles. Wheel games had long been popular, too, where customers put

Crafts, as in 1907, remained a popular part of the State Fair even as new marketing strategies developed in the 1960s and 1970s (State Fair Archives).

a dime on a color and waited for the wheel to stop. Any coin on the winning color received a prize. By the 1970s the line of carnival games extended from the midway to the street between the Kitchens of America and the Modern Living buildings, and included contests ranging from basketball hoops and baseball pitching to darts and ring tosses. Most were privately owned, run by subcontractors who paid a fee and percentage of gross revenues.[10]

Just as varied but without the barkers was food service at the fair. Traditionally, space had been rented to various vendors who offered everything from beans and cornbread to onion burgers and cafeteria food. Church groups and nonprofit organizations had the most popular sites, especially the Hospitality Club, a service group that started serving submarine sandwiches and other treats as early as the 1960s. By the 1990s they had one of the most profitable booths with a featured site in the Made in Oklahoma Building.[11]

Standard fair items such as corn dogs and specialty foods such as cinnamon rolls were provided by private vendors who followed the "fair trail." In the early days of the fair, they had come with portable tents, called cook houses, and gas-fired grills called "flashes." Eventually, most adapted new technologies and traveled with self-contained trailers and full kitchens. One of the more successful vendors at the State Fair of Oklahoma was Phil Little, a native of Minnesota who had started in the

carnival and fair business in the 1890s selling lemonade
out of a bowl. By the 1930s, when the State Fair of Oklahoma was one of his major stops, he had several cook
houses, including hamburger stands and a full-service
cafeteria.

Little employed a man and wife team who would
become fixtures at the fair, Fred and Louise O'Neal.
Both were from Minnesota, and they traveled with Little
along the fair trail that included one-week runs in
Sedalia, Missouri, Des Moines, Iowa, Oklahoma City,
Tulsa, Muskogee, and Shreveport, Louisiana.

Working for $1.50 a day, Louise operated the cash
register in the cafeteria. In 1947, while set up at the
Shreveport fair, Louise noticed a local vendor selling a
new item called a corn dog. As she later recalled, "That's
all the people were eating." She rushed to the kitchen
and worked up some pancake dough with corn meal and
began experimenting with her own version of a corn dog.
Within the year, her new treat was the biggest seller at all
fairs, especially in the South. Eventually, Louise and her
husband developed their own recipe and started selling

*Television personality Ida B (right)
won the pie-eating contest in the
Hobbies and Crafts Building in 1967
(State Fair Archives).*

Dean A. McGee served as president of the fair board from 1971 to 1977, a period of continued development and expansion of programs (State Fair Archives).

Automobile racing reached new levels of popularity from the 1950s to the 1970s so the State Fair offered a wide variety of stocks, modified, super modified, and midgets, seen here (State Fair Archives).

to other vendors. Sold under the name Pronto Pup, the O'Neal's corn dog would become a staple at the State Fair of Oklahoma.[12]

Private vendors such as the O'Neals never would have invested in bigger and better facilities had it not been for a growing attendance at the fair. And of all the enticements the fair could offer the general public, none was as effective at getting people through the turnstiles as entertainment.

The biggest impact on entertainment at the fair after 1964 was the construction of the Arena, which allowed the fair board to book bigger and better shows. Previously, staged events had been held outdoors in front of the Grandstand, where inclement weather always posed a potential problem and the track separated performers from the audience. The former problem had become all too apparent one fall when a heat wave arrived about the same time as an outdoor ice skating show. The result was

a cramped, slushy stage with skaters plowing through an inch of water.

Moving quickly to take advantage of the new Arena in 1965, the fair board was faced with two options. They could bring in the Hubert Castle Circus, which would generate potential profits of $15,000, or book a new show called the "Ice Capades," which would bring in about $10,000. The board chose the latter even though it would generate less profit. As one board member explained, "The Ice Capades would improve the entertainment image of the fair." It turned out to be the right decision.[13]

Typical of any new undertaking, adapting the Arena for an ice skating show posed unique problems. Because the Arena had no built-in equipment, the fair rented 250 tons of ice-making equipment, the equivalent of 900 household refrigerators. The process started with a layer of sawdust on the arena floor on top of which workers installed pipes connected to the compressors. A brine solution was pumped into the pipes and cooled to 16 degrees. To make the ice, workers sprayed water over the pipes for three days, building up layer after layer until the ice was three inches thick. During the performances, the ice was resurfaced and sprayed three times a day.[14]

The 1965 edition of the Ice Capades was titled "Your Grand Tour," with international musical numbers featur-

Beginning with "Auto Polo" before World War I, the State Fair offered a steady venue for automobile daredevils. This line formed for a show in 1959 (State Fair Archives).

ing New York's Park Avenue, Japan's "Night of the 13th Moon," a Paris street scene in 1890, a football game, and a Spanish style rendition of "Bolero." A company of 100 skaters led by German skating star Inge Paul kept the pace fast and furious as sets and costumes changed between each number. Through seven performances, the paid admission was 45,073 with gross ticket sales of $132,800, numbers that convinced the fair board to invest $100,000 in permanent ice-making equipment. For the next twenty years the Ice Capades would remain a fixture at the State Fair of Oklahoma.[15]

While the Arena had to be adapted to ice shows, it was built for rodeo, which earned even more money for the fair. The first rodeos had been staged at the State Fair of Oklahoma in the 1930s, and they had been popular attractions ever since. Rodeos and horse shows were so entrenched in the program that when plans were considered for the Arena, city officials sought the approval of the Rodeo Cowboys' Association and the American Horse Show Association. Both groups granted their blessings. When the Arena opened, it was billed as the only indoor arena in the country with permanent bucking chutes and a 130-foot tunnel running to outdoor corrals so stock could be conveniently led to the chutes.[16]

Beginning a long-term arrangement, the first State Fair Rodeo in the Arena was produced by Beutler and Son, a family-owned company based in Oklahoma that had been one of the top rodeo stock firms in the nation since the 1920s. For the first rodeo in the Arena, Beutler

and Son provided 230 head of stock, including 45 "big
and mean" Brangus roping calves that weighed 190
pounds apiece and 35 steers with "wide, sharp horns" for
the doggers. Clem McSpadden, a great-nephew of Will
Rogers and at that time President Pro-Tempore of the
Oklahoma Senate, served as the first announcer as cowboys competed for $11,000 in prize money. That first
year, top hands included Larry Mahan, all-around champion several years in a row, and Shawn Davis, the premier saddle bronc rider in the nation.[17]

Entertainment, from the midway to the rodeo, played
an important role in the development of the fair. Each
attraction lured crowds who wanted to have fun which, in
turn, generated revenue to support programs that were
not income producers, such as premium competition and
educational displays. Why not, asked Gaylord and the
fair board, use this same combination of entertainment
and revenue production in other areas off the midway
and outside the Arena. It did not take long to test the
possibilities.

*Six of the 32 "Ice Capets" who per-
formed synchronized dances for the Ice
Capades (State Fair Archives).*

In December of 1963 Saunders asked the fair board to consider a new attraction that was a combination ride and practical transportation system: a monorail. Saunders predicted that more than 150,000 people would pay to ride an elevated train during the fair, generating enough revenue in just two years to pay for the entire investment. He had seen a monorail at Disneyland, and he knew that it would not only become a major entertainment attraction, but also a distinctive addition that would enhance the fair's image as a source of progress and innovation.

Members of the board knew the fair could not afford a multi-million dollar system like the one at Disneyland, so they searched for a cost-effective alternative. They found it in Wildwood, New Jersey, where a firm called Universal Designs had built a monorail system for much less. With approval from the fair board, Saunders signed a contract for a prototype train in March of 1964. Two months later, he announced to the public that the fair was going to build the "longest monorail in the Western Hemisphere."[18]

The track, more than a mile and a quarter long, started at the Grandstand and made a wide loop north to the Science and Arts Museum, ran east and north to the Arts Center, then continued south to the 4-H and FFA buildings. It passed over the OPUBCO Building, in front of the Agricultural Building, down the median to the livestock complex, and south and east along Grand Boulevard back to the Grandstand. Depending on the terrain, the track ran from 22 to 25 feet overhead and was constructed with 109 60-foot concrete horizontal beams resting on reinforced concrete posts 55-feet long. For stability, each post was anchored in a massive slab of concrete.[19]

Each train consisted of an engine and four passenger cars with seating for three people abreast. Sleek, stainless steel bodies, aerodynamic lines, and automatic doors evoked an image of space-age technological wonder, just as the board had predicted. More amazingly, the cost of the entire system, including track and three trains, was only $298,000 and went from design to completion in only three months. By opening day of the 1964 fair, the monorail was ready for its first paying customers.[20]

Unfortunately, the visual success of the monorail was not matched by its initial performance. The trains, powered by an innovative hydraulic system, did not work properly. Oil splattered people below the track, while the underpowered engines could not pull a full load. The staff had to remove one car from each train and cut the number of seats per trip from 50 to 25, causing costly delays and long lines. After the fair ended, the fair

Completed in 1964, the monorail became a symbol for the State Fair of Oklahoma (Courtesy Daily Oklahoman*).*

Opposite page: Assembling the monorail trains became an annual sign that the fair was about to begin (Courtesy Daily Oklahoman*).*

board sent the trains back to New Jersey where Universal Designs installed electric motors at a cost of $100,000. By the next year, the system was working as planned and making a steady profit for the fair at 75 cents a rider. Years later, Gaylord and Saunders would judge the monorail one of the best investments ever made at the fair.[21]

Another project that shared the twin benefits of entertainment and revenue production was the space tower. As early as 1962 Saunders had reported to the fair board that the most popular feature at the Seattle World's Fair had been a space needle. The idea was resurrected in 1967 when the old Arrows to Atoms tower, constructed in 1957 to commemorate the state's semicentennial birthday, was condemned as a safety hazard. When it was learned that a tower in Minnesota grossed $89,000 the first year of operation and $62,000 the second, the fair board approved plans to proceed with an Oklahoma version of a space needle.[22]

Completed at a cost of $450,000, the new Arrows to Atoms Space Tower was patterned after one recently built for Expo '67 in Montreal. It was 330 feet tall and anchored 50 feet underground in a base of 700 cubic yards of concrete. Outside, a doughnut-shaped rotating elevator carried as many as 60 passengers on each ascent. Despite being out of commission for a day and a half for repairs at the 1968 fair, the attraction carried 48,455 paying customers.[23]

Investments in new attractions continued into the 1970s as the fair board responded to new trends and market demands. By this time the board was led by Dean A. McGee, chief executive officer of Kerr-McGee Industries, a global energy firm based in Oklahoma City. Under McGee's leadership from 1971 to 1977, the fair board continued the expansion program begun during Gaylord's ten-year tenure as president.

In 1976, as a result of growing interest in gardens and urban landscaping, the fair board funded a 16,800-square-foot addition to the Made in Oklahoma Building for an expanded Garden and Flower Show. The central feature of the exhibit was a lush garden designed by landscape architect Kimio Kimura, who used hundreds of plants, walkways, bridges, fountains, and pools to create a colorful, pastoral scene. Two years later, with that success to build on, the fair board constructed the new Garden and Flowers Building for $525,000. Six different landscape areas filled the 26,000-square-foot building. There was an Alpine mountain scene with a 20-foot waterfall, a western frontier village, and a tiny hillside chapel. Also included was a flower arrangement 16 feet tall, an 18-foot-tall figure sculpted from a redwood log, and a 26-foot-long wagon train. Competitions for plants and flowers filled out the new building.[24]

Another response to public interest, this time the historic preservation movement, was the Goodholm Mansion, a three-story Victorian home moved to the fairgrounds in 1979 after it was condemned to make way for a new highway on Oklahoma City's east side. Reconstructed on the grounds at an initial cost of $131,000, the

Indian dance performances were a feature of the State Fair as early as the 1930s. These fancy dancers were part of the 1978 fair (State Fair Archives).

mansion was adopted by the Women's Committee of the Oklahoma City Symphony as the Decorator Show House in 1980. Thereafter, the home was steadily renovated to "show how people lived at the turn of the century." By 1990 it was completed and open to the public, filled with furnishings from various antique dealers and managed by hostesses in period dress provided through the Oklahoma Society of the Daughters of the American Revolution.[25]

The close relationship with outside groups such as the DAR was typical of the fair's growing reliance on partnerships, especially when cooperation drew crowds, generated income, and fulfilled the basic mission of the fair to provide all things for all people. But of all the partner-

ships forged over the years, none was stronger or more natural that the cooperation between the fair and the Oklahoma City Chamber of Commerce.

It was a natural alliance, easily understood. The fair, as an institution, combined the objectives of education, entertainment, and economic development into one cohesive program. Chamber officials, dedicated to the economic development of Oklahoma City, saw the fair as an effective ally as they launched a host of new programs.

A good example of the growing partnership followed on the heels of World War II when the Chamber increased its emphasis on labor-intensive industrial manufacturing as a way to diversify Oklahoma City's economy, create jobs, and stimulate investment and production. This was accomplished through industrial parks, recruitment of international companies, and financing for large plants and sophisticated equipment. Another effective tool in this campaign was a trade show, first staged in 1946 at the Municipal Auditorium, that featured Oklahoma companies and their products. It was an easy leap to adapt the "Made in Oklahoma" theme to the state fair.

Although Oklahoma companies had long supported

the fair, there had never been a featured exhibit area that focused specifically on manufacturers with Oklahoma roots. In 1963, for the first time, the fair dedicated part of the Appliance Building to a "Made in Oklahoma" exhibit. The effort was successful enough that it was repeated and expanded several years in a row. In 1965 the exhibit was tripled in size, boosted by the addition of three new awards sponsored by the Chamber. One category was in "Outstanding Products," a distinction given those first few years to Southwest Electric Company, Delkids, Inc., Sequoyah Mills, and Frankoma Pottery.[26]

In 1969, carefully watching the public's reaction, the fair board reallocated space so the Made in Oklahoma exhibit could expand. Initially, to make more room, the international exhibit was moved to the Appliance Building, and the old International Building was renamed the Made in Oklahoma Building. Three years later, prompted by the need for even more space, the Made in Oklahoma exhibit was moved into the old Agricultural Building, which was remodeled, expanded, and air conditioned. There, it would continue to grow, providing a showcase for Oklahoma firms.[27]

The fair's efforts to promote Oklahoma's economic development did not stop at the state borders. In 1966, looking for new exhibitors and attractions, the fair board sent Saunders on a recruiting mission to Europe, where he met with government officials and businessmen who

Lieutenant Governor George Nigh, as an advocate of tourism, used the State Fair as an outlet for promoting Oklahoma parks and recreation in 1971 (State Fair Archives).

might send exhibits to Oklahoma City. As stated in the annual report that year, the objective was to "develop markets for the sale of foreign products...and to give Oklahoma manufacturers the opportunity to make personal contact with foreign businessmen." Gaylord and the fair board also saw the international exhibit as a way to attract crowds and generate desperately needed revenue.[28]

In addition to the Oklahoma City Chamber of Commerce, the fair had the support of the United States Department of Commerce, which granted "trade fair" status and provided options for duty-free importation of goods. Also lining up behind the program was the United States Department of State, attracted initially through the office of Congressman Happy Camp. For several years in a row, officials from the Department of State sent brochures and cover letters to select ambassadors touting the advantages of the Oklahoma City fair. When Saunders and his salesmen arrived in foreign countries, they had the advantage of "going through diplomatic channels."[29]

Over a ten-year period, staff members made official recruiting visits to 196 countries. Typically, they would

Like midway rides, stunts staged by the Joie Chitwood Thrill Shows got bigger and more flamboyant by the 1960s (Courtesy Daily Oklahoman*).*

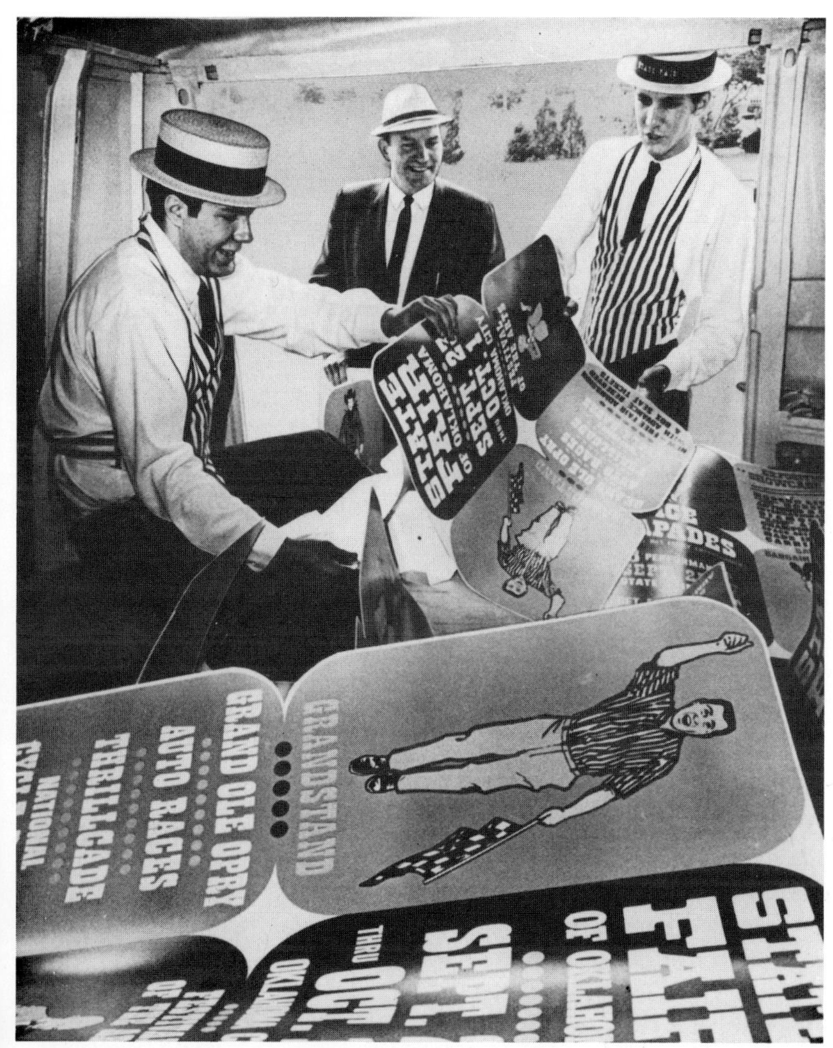

College students Ken Strange and Greg Lowry loaded 500 promotional signs for delivery as Sandy Saunders looked on (Courtesy Daily Oklahoman).

seek appointments with officials highly placed in government. "If we couldn't get appointments with someone who had the authority to say yes or no, we didn't go," Saunders recalled. That policy made for efficient trips. "One year we were visiting fifteen countries on one trip...and we stayed only two or three days in any one country," he remembered. Still, he occasionally encountered problems. In Kenya, he met with the mayor of Nairobi, Elizabeth Kinyata, who was interested in the possibilities of greater trade with the United States. When militant nationalists on her staff tried to disrupt the negotiations, the mayor "got up and carved them a new one in a language they understood," recalled Saunders. Kenya sent an exhibit to Oklahoma City the next year.[30]

In 1967, the first year of the expanded International Exhibit, thirteen foreign nations sent delegates and displays, including official government participation by Finland, France, Japan, Mexico, Republic of China, and Sweden. Commercial exhibitors came from the Bahamas, Germany, Hong Kong, India, Philippines, Korea, and

Switzerland. Notable exhibits were staged by Israel, whose slogan was "The Bible is the guidebook," and Japan, whose trade delegates assembled a 2,000-square-foot display that included a tea ceremony reenacted several times each day.[31]

Personal recruitment continued to pay dividends. In the depths of the Cold War, board members worked through the Institute of Soviet-American Relations to get Saunders an appointment with the mayor of Moscow. At the meeting the mayor got right to the point and asked why a fair in middle America wanted to host a Soviet display. Saunders, equally direct, responded "we have got to learn to get along with each other, and you have got to learn what Americans are like and Americans need to know what the Russians are like....Only then can we come to any kind of agreement where we can keep peace in the family." The Soviets, obviously impressed with such sincerity, sent an exhibit to Oklahoma City the next year.[32]

The Soviet Union was the headliner in 1971 with an extensive exhibit on everyday life in Russia. Through photographs and ambassadors of good will, they tried to explain the differences in the two countries. That same year a commercial exhibitor from Mali assembled a show of art objects made by the Dogon tribe in the 18th and 19th centuries. The dealer also sold modern items such as quilts, hand-woven blankets, and tie-dyed fabrics.[33]

Foreign participation peaked in 1971 when 23 foreign countries sent exhibits to Oklahoma City, an impressive number considering the fact that all Asian countries

cancelled at the last minute when President Richard
Nixon took steps to cut America's trade deficit through
protectionism. Such success helped convince the fair
board to initiate the first International Trade Sympo-
sium, an annual event co-sponsored with the Oklahoma
City Chamber of Commerce. Featured speakers over the
next few years included the deputy director of International Trade with the U.S. Department of Commerce and

Even before parimutuel betting was legalized, the State Fair maintained the tradition of horse racing. This harness race was staged at the 1976 fair (Courtesy Daily Oklahoman).

the senior advisor to the President of the Inter-American
Development Bank. Both specialists saw great potential
for the sale of Oklahoma oil, gas, aircraft, and agricultural technology throughout Latin America.[34]

In 1984, to house the ever-popular international
exhibit, the fair board invested $2.3 million, the largest
single sum in the fair's history, to construct the International Trade Center Building. It was a distinctive, innovative structure with an air-supported dome roof and
70,000 square feet of space. That first year, the white,
domed structure featured an elaborate exhibit from the
People's Republic of China, which took up 10,000 square
feet of space, followed the next year by an even bigger
display from Kyoto, Japan, that included displays of
native arts and crafts, travel, and food. A stage provided
the setting for kimono fashion shows, dances, and tea
ceremonies.[35]

Much of the credit for the success of the international
exhibit went to the Sundowners, a volunteer group organized to host foreign visitors. The tradition for that
volunteer help had started with the State Fair Committee
of the Oklahoma City Chamber of Commerce, whose
members for years had served a variety of volunteer
functions from hosting livestock shows to ticket taking.
In 1965, for the first time, the Chamber set up an office

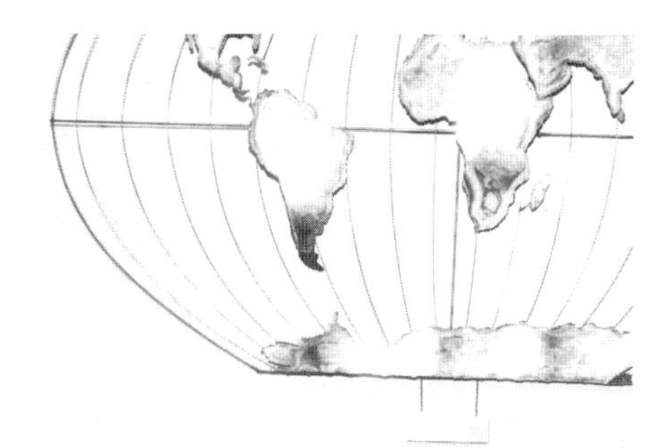

Sundowner Dorothy Newcomb (left) talks with Ann Banerji of India (State Fair Archives).

Bill Swisher played the role of Mr. Oklahoma Territory in a mock wedding ceremony for Statehood celebration (State Fair Archives).

on the fairgrounds with volunteer coordinators. When the international exhibit was launched, the International Visitors Council of the Chamber stepped forward to host visiting families.[36]

After several years of increased involvement, members of the two Chamber committees decided to organize a special unit to "offer the hand of friendship and service to visiting dignitaries and exhibitors." In 1972 they organized under the name "Sundowners," with 33 charter members. To promote teamwork and equality, they decided that every member would be a vice-president. To provide distinction, they adopted an official uniform consisting of a white blazer and dark shirt.[37]

As the Sundowners became more proficient, they expanded their role to organizing meetings, conducting tours, and planning receptions. "Our purpose is to keep

exhibitors happy," said one member, "because we want them to keep coming back." By 1979 the Sundowners hosted two lounges, one in the International Building, the other in the Travel and Transportation Building, where 160 members served the needs of all exhibitors, not just international guests.[38]

The Sundowners continued expanding their role until it became a year-round operation. After 1980 they assumed more responsibility for interim events, such as the Auto Show held every spring, and the horse shows which started in the fall and ran most of the year. Their duties also expanded during the fair. They staffed information booths, coordinated daily flag ceremonies, hosted several lounges for exhibitors, and filled in whenever and wherever their talents and energy were needed. True to their credo, they "offered the hand of friendship and service."[39]

A miniature chapel from a frontier town was part of the new Gardens and Flowers Building, opened to the public in 1978 (Courtesy Daily Oklahoman).

With allies such as the Sundowners and the Oklahoma City Chamber of Commerce, the fair grew faster than anyone had anticipated, providing "all things to all people." The proof was in the numbers. In 1980, after 20 years under the new management system set up by Gaylord and Saunders, the fair attracted 1,336,409 paid admissions, an impressive leap from the 455,671 people who had attended the fair in 1960. Even more dramatic was the increase in revenues. In 1960 gross income was $328,671; by 1980 it was $2,931,252, a cash flow that not only made the fair a self-supporting enterprise but also provided funds for improvements.[40]

By every measure, the fair had turned into a giant success, ready for a promising future. Only one persistent problem remained to be solved—the division of management control over the fairgrounds.

Chapter Eight
EXPANDING HORIZONS

From 1961 to 1980 the State Fair of Oklahoma exceeded the most optimistic expectations of the fair board. Attendance broke records year after year. Attractions and exhibits improved and expanded. And revenues rose dramatically, making it possible for the fair board to invest $7 million in capital improvements. There was only one major blemish on the record, one problem that had not been solved—the division of year-round authority over the fairgrounds.

City control of the fairgrounds eleven months out of the year had been forced on the fair board in 1957, a time when the fair was deep in debt and overwhelmed by the demands of a rough, sprawling new site. Thereafter, the State Fair of Oklahoma was treated like any other tenant with the special provision that it had exclusive authority over the entire grounds one month out of the year.

Opposite page: John Keel, a member of the Plains Indians Dance Troupe, performed at the Cottonwood Post Stockade in 1993 (Courtesy Daily Oklahoman).

The split system had many disadvantages, but the most frustrating was city politics with its cross currents of unrelated issues, factionalism, and tendency to deal with immediate solutions at the expense of long-term needs. As early as 1961 the new chairman of the executive committee, Edward L. Gaylord, complained about the damaging effects of politics when the city council balked at a $400,000 gift from philanthropists John and Eleanor Kirkpatrick. Gaylord publicly stated that "This proves the necessity for returning year round control of the fairgrounds to the State Fair Association and removing supervision and control of the fairgrounds from city hall politics." Five years later, he again bristled at political meddling, claiming that "harassment from city hall has retarded growth at the fair."[1]

The cooking department, seen here in 1989, is a tradition that dates to the first State Fair of Oklahoma in 1907 (Courtesy Daily Oklahoman).

While Gaylord and other board members recognized the political failings of split management, Saunders and the staff witnessed the inefficiency on a daily basis. Never one to mince words, Saunders regularly complained about sloppy maintenance, disregard for long-term needs, and inconsistent policies for use of facilities. In truth, Saunders' aggressive, sometimes abrasive style aggravated the problems at times, causing an escalation of conflict and breakdown of communication. Individual council members, sympathetic city employees, and even the leadership at the Chamber of Commerce occasionally tried to heal the rift, but in large part, little was done about the basic structural problems other than promises from city hall to do a better job.[2]

In 1978 leaders on the fair board decided it was time

"Toe Tappin Entertainment" was the theme at the 1982 State Fair of Oklahoma, a theme demonstrated daily at the OPUBCO Pavilion (State Fair Archives).

John Parsons, president of the fair board from 1978 to 1981, began the negotiations that successfully returned control of interim operations to the fair board (State Fair Archives).

at last for a final resolution. The chairman of the board was John Parsons, president of the Oklahoma Division of Southwestern Bell and an active civic leader who had his finger on the pulse of the community. With the help of several board members, Parsons initiated negotiations with Mayor Patience Latting, the city manager, and select city council members, while the fair board's legal counsel, Joe Rucks, discussed the legal technicalities of a new operating agreement with Walter Powell, municipal counselor for the city.[3]

From a strictly financial point of view, transferring control of the fairgrounds to the fair board would benefit the city. According to City Manager Jim Cook, the city had spent $1,045,586 during the previous year subsidizing interim operations at the fairgrounds. In his opinion, even if the city made a cash contribution of $550,000

to the fair board and furnished utilities, the city would still save money if interim operations were transferred to the fair board.[4]

Even with the potential of this financial advantage, as well as a recommendation from the city manager to negotiate a new contract, the process quickly ran into trouble. The most serious problem was the basic question of whether or not the city could grant control of public property to a private corporation. As a solution, Rucks suggested the formation of a public trust. Through such a public entity, approved by the city council, the fairgrounds could be leased for year-round operation. Even though the municipal counselor agreed that such a procedure was legal, the city council failed to act.[5]

Parsons reported at the 1978 annual meeting that two more problems had disrupted negotiations. One was Walter Powell's demand that a new agreement include a "whim clause" allowing the city the right of arbitrary, unilateral cancellation of the agreement without cause at any time. The other stumbling block was a demand that the fair board comply with the open meeting act of the State of Oklahoma. Mayor Latting reported that three members of the council would approve the contract as written, but the rest would vote no unless the fair board opened its meetings to the public.[6]

The issue was taken up again in 1980. Leading the fair board's negotiating team were John Parsons, chair-

Free educational demonstrations, such as this lecture by the American Heart Association, carry on the longtime traditions of education at the fair (State Fair Archives).

William Hulsey, chief executive officer of Macklanburg Duncan, was elected to serve on the fair board in 1970 (State Fair Archives).

man; Bill Hulsey, 1st vice-chairman; Bill Swisher, 2nd vice-chairman; Charles Vose, treasurer; Sandy Saunders, president; and Frances Young, secretary. Appointed to the executive committee to serve along with the officers were Edward L. Gaylord, Dean A. McGee, and Paul Strasbaugh.[7]

In early May Parsons reported to the fair board that an agreement had been reached between Walter Powell and Joe Rucks. If approved by the city council, it could be effective as early as July 1. Although there was overwhelming support in the community to have the city council take this step, there was one member who opposed the new contract, Jerry Gilbert, longtime opponent of the state fair. Gilbert stated that the nine-member city council could, in his opinion, manage the fairgrounds more efficiently than a thirty-member fair board.[8]

Link Carnival provides a wide variety of rides for both thrill seekers and small children (State Fair Archives).

A 4-H student grooms her steer for competition (State Fair Archives).

Despite that opposition, the new agreement was approved by the council and signed on July 1, 1980. The actual parties to the contract were the fair board and the Oklahoma City Public Property Authority (OCPPA), whose trustees were the mayor and members of the council. In reality, the agreement was between the city and the fair board. OCPPA still had some outstanding revenue bonds issued for improvements to the Arena and 89er Stadium, so the fair board agreed to cover debt service on the revenue bonds until July, 1991.

Under the terms of the agreement, the city removed its personnel from the fairgrounds, but turned all public events equipment over to the fair board, which then became responsible for maintenance, replacement, and additions. The city also agreed to provide all utilities except for the period of exclusive use when the fall fair was conducted. The fair board was designated by the city as its "interim operating agent." For this service the city agreed to pay the state fair an annual management fee of $300,000. The term of the lease was 25 years, and for the first time the "whim clause" was included in the agreement.[9]

With the successful completion of negotiations, Parsons announced his desire to step down as chairman of the board. A longtime civic leader, he stated there were

Continuing a tradition that began in 1892, the State Fair of Oklahoma recognizes excellence in domestic skills (State Fair Archives).

other matters he wanted to accomplish. His desire was agreed to with regret. The new chairman was Bill Hulsey, chairman and chief operating officer of the Macklanburg-Duncan Company. His late father-in-law was Lou Macklanburg, president of the fair board from 1951 to 1954 when the new fairgrounds had become a reality. Hulsey was a strong advocate of the state fair and its programs and an outstanding civic leader involved in many leading efforts, including the Chamber of Commerce, the Cowboy Hall of Fame, the Sirloin Club, and the Southwest American Livestock Foundation. These latter two organizations, products of the Chamber of Commerce, sponsored the largest junior livestock show in the nation each year at the Oklahoma City fairgrounds.

Unfortunately for Hulsey, the operating agreement signed in 1980 was destined to be short-lived. The ink was hardly dry on the document when grumbling started at city hall. Saunders reported at the annual meeting that the city wanted to renegotiate the agreement covering interim operations, so the fair board authorized the executive committee to deal with the situation or move the state fair to a new location. Councilman Robert M. Frank, in a television interview, flatly stated that the

The Junior Hospitality Club began
hosting a food booth in the 1960s, a
popular tradition that would continue to
the present time (*Courtesy* Daily
Oklahoman).

council was about to cancel the fair board's 25-year contract to operate the State Fair of Oklahoma.[10]

Frank, however, did not represent a majority on the
city council. Councilmen Jack Cornett, Goree James, Jim
Scott, and Bill Bishop realized the value of the state fair
to Oklahoma City, a perception that had been strengthened through a public information campaign that included a video presentation on the goals and programs of
the fair. Frank and Gilbert would remain critical of the
state fair throughout their tenures on the city council,
but they would never achieve a majority.

Robinson's Racing Pigs round the track for a cookie (Courtesy Daily Oklahoman).

Most of their criticism against the fair board focused on accusations of foul play and complaints about a fair board policy regulating food and beverages sold by concessionaires during the fair. The most widely publicized claim was made by out-of-town carnival operators Gene Sorrows and Prier Price. Escorted in by Councilman Frank, they asked the city council to investigate contracts and arrangements between the State Fair and Rod Link's Amusement Corporation of America. Sorrows, who claimed to represent Expo Shows, which had unsuccessfully sought to play the Oklahoma City fair, claimed that the fair board and staff members were receiving monies under the table to maintain the contract with Link. Saunders responded that the contract with Link was openly negotiated and that no monies exchanged hands other than what was specified in the contract. Quickly after making his claims, Sorrows left town, never to be heard from again. None of his charges were ever substantiated.[11]

A more substantive complaint was the new policy regulating food vendors. After the city transferred interim operations, the fair board reserved the right to approve qualified vendors as a means of controlling quality and safety of products on the fairgrounds. Certain concessionaires sought to break this practice and select their own vendors. Saunders and the board knew that once this was done, there would be no control, and management would lose one of the key strengths on which it had built a worldwide reputation.[12]

The other controversial issue was the right to set rental rates on buildings. When the city controlled interim operations, rental rates varied widely, mostly for political considerations. Even the Chamber of Commerce brought pressure to secure lower rates for events it sponsored. When the fair board took over interim operations, the city's subsidy no longer provided a financial cushion for such discounts. Needing steady, reliable revenues from interim users, the fair board adopted a policy of using only published and approved rates.[13]

At the beginning of 1982 these nagging issues between the city and fair board were still not resolved, so Saunders reported to the board that all improvements were being held pending a new agreement. Then came the untimely death of Hulsey, who had been ill for several months. Fortunately, the leadership role was picked up by Bill Swisher, 1st vice-president, who had founded CMI, one of the largest construction machinery manufacturers in the country and a man who had come to the aid of the fair several times in the past. Swisher met five times with the parks committee of the city council, which was chaired by ward 3 councilman Jack Cornett, then

The Midway peaked as a destination favorite in the 1960s (State Fair Archives).

reported that most of the issues on the new contract had been resolved.[14]

Approved by the fair board on August 3, 1982, the contract included a 25-year lease, payment on OCPPA bonds, a whim clause, a management fee that would be reduced 20 percent per annum over a five-year period, and utility payments that would be reduced 20 percent per annum for five years until the full amount of utility costs had been paid by the fair board. To take advantage of lower rates, the city agreed to maintain hazard insurance covering all improvements on the fairgrounds. The fair board was required to carry workmen's compensation and hazard insurance.[15]

The new mayor by this time was Andy Coats, a former district attorney who was well aware of Frank's and

The outdoor display of aircraft pays tribute to the role that aviation has played in the economic and social development of Oklahoma (State Fair Archives).

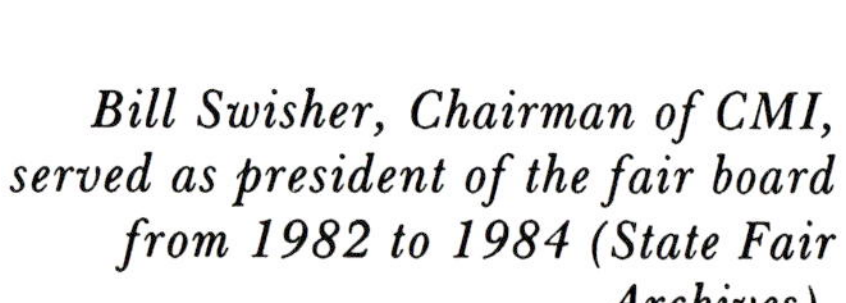

Bill Swisher, Chairman of CMI, served as president of the fair board from 1982 to 1984 (State Fair Archives).

Gilbert's hostility towards the state fair. Supportive of the new agreement, he named seven members to the new Fairgrounds Trust, including himself, two members of the council, and four people from the general public. The mayor, according to the agreement, could appoint a surrogate trustee to act in his absence, while the city manager could attend all trust meetings as an advisor.

No sooner had the fairgrounds trust been accepted by the city and the leases assigned than councilman Frank voiced objection to the makeup of the trust. He notified the mayor that he would propose changes at the next meeting of the council unless another councilman was added to the list of trustees and the city manager given a vote on trust matters. He contended that the city should have voting control since it owned the facility. He also notified Mayor Coats that he was prepared to circulate a petition calling on the U.S. Department of Commerce to deny a world's fair application when it was submitted.[16]

On October 3, 1983, after Frank admitted he did not have the votes to block the transaction, the city council adopted the trust indenture of the Oklahoma City Fairgrounds Trust and leased the fairgrounds to the trust with the exception of the baseball stadium. In addition, it assigned all existing leases on the fairgrounds to the trust. These leases covered OCPPA, Boy Scouts Headquarters, Oklahoma Art Center, Oklahoma Science and Arts Foundation, and the State Fair of Oklahoma. At long last, management of the fairgrounds was consolidated under one board and removed from the volatile crosscurrents of city politics.[17]

With that issue behind them, fair board members turned their full attention to yet another new challenge, staging a world's fair. The international extravaganza was to be the central event of the centennial of the land run of 1889, which marked the first non-Indian settlement of the state of Oklahoma and the 100th birthday of the founding of Oklahoma City. To city leaders, a world's fair would be the perfect capstone to draw attention to a century of progress.

Planning for the city-wide event was consolidated under the combined leadership of the Economic Development Foundation, the fair board, the Chamber of Commerce, the city, and the trustees of the Fairgrounds Trust. Committee members representing these special interest groups included G.T. Blankenship, chairman, Goree James, vice-chairman, Paul B. Strasbaugh, secretary, and members Andy Coats, Jack Cornett, Jim Daniels, Lee Allan Smith, Bob McCoy, and Scott Johnson, city manager. In a show of support, the fair board committed $90,000 a year for two years to the Oklahoma City Economic Development Foundation to help finance work on the application, a lengthy process that would start with its submission to the United States Department of Commerce in Washington, D.C., and if accepted and

John Sweifel hand carved this 60-foot replica of the White House, which was on display at the fair in 1991 (Courtesy Daily Oklahoman).

approved, go to the Bureau of International Expositions, an International Treaty Organization located in Paris, France.[18]

Paul Strasbaugh, executive vice-president of the Economic Development Foundation, told the fair board that the application would take two years to develop and include a master plan, financial feasibility, theme, legal programming, promotions and advertising, and a complete plan of operations and staffing. He predicted that construction of the necessary buildings on the fairgrounds alone would cost $50 to $60 million. If another site was used, the cost would run to $100 million or more.[19]

Plans proceeded rapidly. John Montgomery, a Washington, D.C., consultant, was retained to represent the planning committee at the Department of Commerce and the Oklahoma congressional delegation. Dr. George Pratt, in charge of International Exhibits at the Department of Commerce, also was consulted.[38] Other specialists recruited through the international program of the 1983 State Fair of Oklahoma included Olle Herrold of Finland, a member of BIE and president of European Trade Fairs Association, King Cole, president of the Spokane Worlds Fair, and Paul Creighton, past opera-

This junior horseman won first place honors in the Appaloosa competition at the 1990 State Fair of Oklahoma (State Fair Archives).

tions manager of the Spokane and Knoxville fairs and at that time manager of the New Orleans World Fair. Each of these consultants examined the operations of the fairgrounds and reported that in their opinion it was the best site on which to develop a world's fair.[20]

The New Orleans Fair, due to open in 1984, became the model for Oklahoma City's plan. At that time the target to submit the application to the Department of Commerce was the spring of 1984. The application, assembled under the direction of Strasbaugh, included 15 sections in great detail as required by the Bureau of International Exhibits. It was 825 pages long, assembled in two volumes, the largest proposal ever produced in Oklahoma City. The theme was "Run of 89—The Land."

The world's fair, for a project started with such high hopes, suffered a premature death. The biggest stumbling block was an economic collapse in Oklahoma from 1982 to 1988 that resurrected memories of the Great Depression. The most visible cause of the economic crash was the sudden drop in oil prices from a high of $44 to a low of $10 a barrel. That blow was followed by a real estate crash that ended in the near collapse of Oklahoma's banking system after the failure of Penn Square Bank. The results were massive layoffs, failed companies, and personal bankruptcies.

The disastrous effect of the economic depression was compounded by bad news from other cities that had hosted world's fairs. First came a report about the Knox-

Major General E.L. "Mike" Massad was president of the fair board when Don Hotz was hired as general manager in 1985 (State Fair Archives).

ville World's Fair which included a major scandal involving its president, Jake Butcher, a leading banker in Tennessee who allegedly had used bank funds to shore up losses at the world's fair and engaged in self-dealing lending practices involving members of his family and close friends. Subsequently, he lost his fortune, and his banks were taken over by the FDIC. Even more psychologically damaging were the results of the New Orleans World's Fair, which was a big cultural success but a financial disaster. By the time it closed in the fall of 1984, the New Orleans fair posted losses of more than $120 million.[21]

Given the reality of the local economy and the disappointing results of the last two world's fairs, it was decided to withdraw the application for a centennial world's fair. This action was taken jointly by the Economic Development Foundation and the State Fair of Oklahoma.[22]

Reeling under the disappointment of the aborted world's fair, the fair board was faced with another abrupt change when Orval O. "Sandy" Saunders announced his retirement early in 1985. For 25 eventful and productive years, he had been a dedicated front-line warrior, pushing and pulling and working with board members and staff to transform a struggling, underdeveloped state fair into a world class exposition, the third largest fair in the country. By the time he stepped down on October 31, 1985, the fair board had invested more than $14 million in new facilities, making it one of the best fairgrounds in the nation.[23]

Clearly, Saunders had been the right man in the right place at the right time in 1961 when conditions called for an aggressive, sometimes combative builder who could push hard and get things done. By 1985 the times and the place had changed. Instead of a half-completed fairgrounds with a budget deficit, the modern fairgrounds had a number of quality facilities used by patrons ranging from rock groups to boat shows. Instead of a confrontational relationship with the city government, the problems of split management had been solved, replaced with a new spirit of cooperation and communication. And instead of managing a staff focused on one exposition that lasted only ten days out of the entire year, the job called for someone who could manage a big, complex business that pulsated with energy 365 days a year. The times called for a master manager, a leader who could assess and organize resources, generate citywide support, and then have the drive to generate action and keep it going. The fair board, on November 19, 1985, announced it had found that new leader—Donald J. Hotz.[24]

Hotz came to the fair with a wealth of experience that

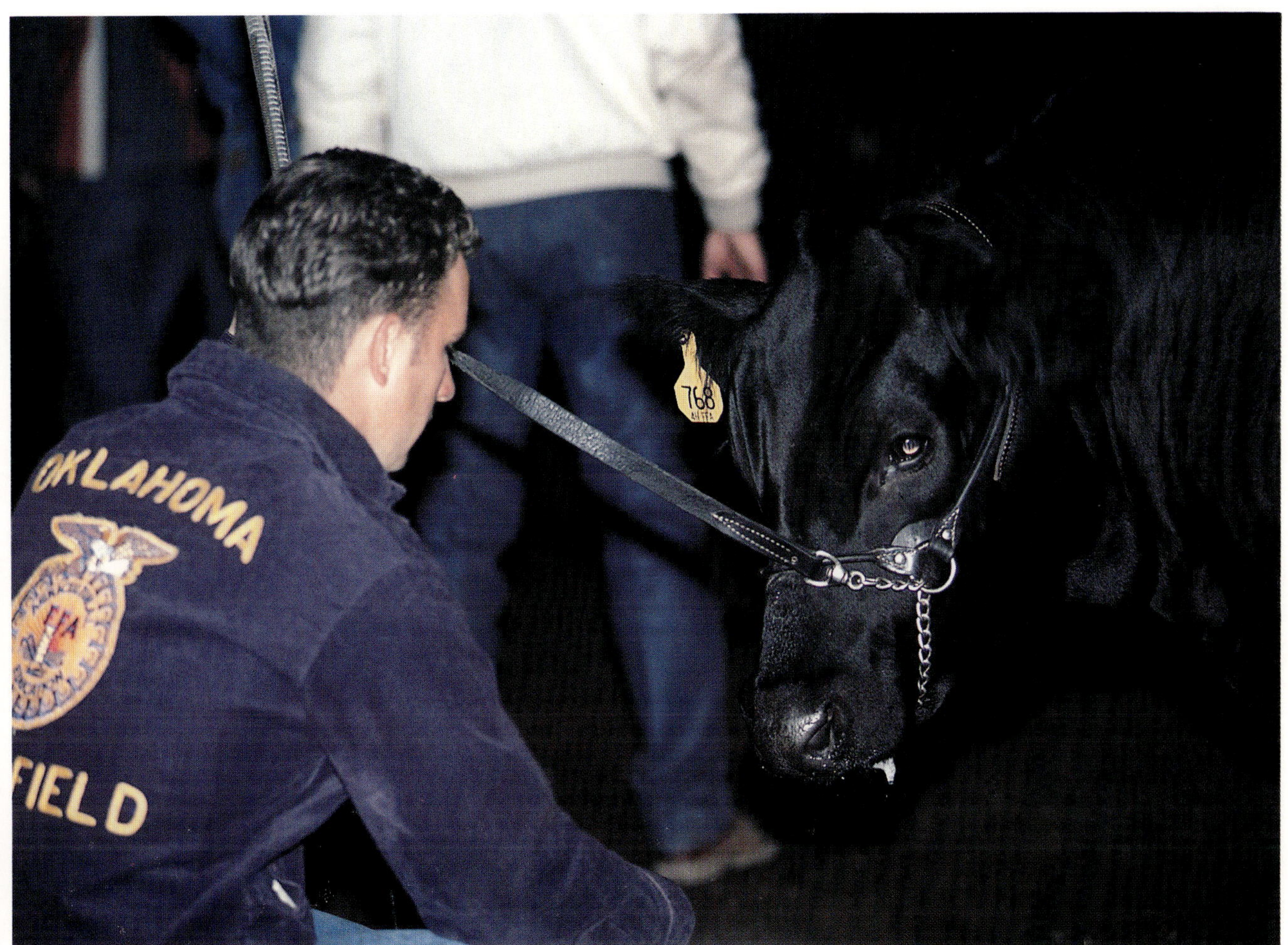

Future Farmers of America have been a central part of the fair's programs since World War I (State Fair Archives).

would prove invaluable. He was a native of western Oklahoma where his father, Elmer Hotz, alternately had operated a wholesale petroleum company, run an automotive repair garage, and built a successful ranching operation. The younger Hotz graduated from Southwestern State University, then taught school for a few years before he signed on with Page Aircraft as a training officer. Following his service in the army, he joined the civil defense operation of the city of Oklahoma City. He then transferred to the public events department where he started his distinguished career in events and facility management.

In 1964 the city named Hotz manager of interim operations at the fairgrounds, which included running the newly constructed State Fair Arena. While at that post, he received a commendation from the fair board, which previously had been critical of the city's management. While he was at the fairgrounds, Hotz was credited with improving facilities and mending relations between Saunders and city officials.

When the city voted to construct the Myriad Convention Center, Hotz was assigned to monitor construction and then to manage that facility, which became the larg-

Edward C. Joullian, III, has served the fair board as chairman during a period of program diversification and a renewed emphasis on traditional strengths (State Fair Archives).

est in Oklahoma and one of the largest in the Southwest. He then had accepted an offer from the University of Oklahoma to supervise construction and manage the Lloyd Noble Center, designed as a multi-use arena.

While he helped build a national reputation for the Lloyd Noble Center as a major sports and entertainment facility, Hotz taught a master's degree level course in facility management. A number of his students would go on to become successful managers in facilities across the nation. During this time he also served as a consultant to universities and cities that were building new sports and entertainment arenas.[25]

When he came on board at the State Fair of Oklahoma, Hotz had the good fortune of working closely with a number of capable board members. In 1985 the chairman of the fair board was E.L. "Mike" Massad, a retired major general who had gained fame as a young man when he earned All-American honors as a football player at the University of Oklahoma. "Iron Mike" served his two-year term in 1985 and 1986 just as Hotz was instituting some important changes in the way the fair was run.[26]

Another board member who worked closely with Hotz was Edward C. Joullian, III, elected chairman of the

Lasers and fireworks were used at opening ceremonies for the 1990 fair (State Fair Archives).

board in 1987. Joullian, like so many members of the
fair board, brought invaluable skills to the management
team. Trained in petroleum engineering, he was president and chairman of the board of Mustang Fuel, one of
the nation's largest natural gas supply and transmission
companies. He also was prominent as a national and
international leader of the Boy Scout movement.

As a Boy Scout in Oklahoma City, he had earned the
rank of Eagle under the encouragement of his father, the
late Edward C. Joullian, Jr., who also was a prominent
Scouter. The younger Joullian would go on to become
president of the Last Frontier Council in Oklahoma,
president of the Boy Scouts of America, and chairman of
the World Scout Foundation headquartered in Geneva,
Switzerland. He also served on the boards of directors
for Colonial Williamsburg Foundation, the Fleming
Companies, the National Cowboy Hall of Fame, and
numerous civic organizations. By 1985, when Hotz was
hired as president and general manager of the fair,
Joullian was ready to dedicate his considerable energies
and talents to the fair's expanded horizons.[27]

Hotz, with guidance from the fair board, had an
immediate impact on the State Fair of Oklahoma. The
first innovation, one that showed his willingness to make
tough decisions, was extension of the fair from ten to
seventeen days.[58] The state fair had operated on a ten-
day basis since 1907, and breaking that tradition was
especially controversial among exhibitors, some of whom
had other commitments both before and after the State
Fair of Oklahoma. Hotz responded that the longer run
was necessary if the fair was to grow and prosper.[28]

The primary reason for the seventeen-day format was
to minimize the effect of bad weather, the traditional
death of any fair. In Oklahoma, where continental
weather patterns usually lasted only a few days, especially
in the fall, a stalled storm system could devastate a ten-
day fair, but cause only a slight shift of attendance over
two weeks and three weekends.

Hotz also recognized that an extended run was
needed if the fair was to accommodate basic changes in
families and schools. Since World War II, more women
had entered the work force every year, limiting the time
available for families to attend the fair during daytime
hours. Too, many schools had traditionally set aside a
"fair day," when students were given free tickets and
school busses were used to transport thousands of kids to
the fair. By the 1960s many districts were forced to
cancel school-sponsored outings as rising travel expenses
and state funding formulas discouraged long distance
excursions. Confronted with these changes in society,
the fair had to respond.[29]

*Donald J. Hotz became President
and General Manager of the fair
in 1985. He and his staff have
adopted new management strategies
that have pushed attendance and
revenue to record heights (State
Fair Archives).*

The addition of five more weekday nights and another weekend made an immediate impact. Attendance jumped from 1,327,799 in 1986 to 1,742,012 in 1987, while revenues increased almost $2 million in one year to $6,117,313, a financial home run that allowed the expansion of existing programs and the addition of new, high visibility attractions.[30]

One of the more popular new features was the sand sculpture project in 1988. Hotz contracted with sculptors Walter Schwartz and Paul Dawkins to create a 300-ton piece of three-dimensional art measuring eight feet tall by 260 feet long. Starting on September 16, before the fair began, they used a mixture of one part cement to twenty parts sand to block out the design. Then they started carving just as the fair began and the crowds started pouring through the Made in Oklahoma Building.

Visitors saw a panorama of history come to life. First came a large dinosaur followed by a cave man, a Viking ship, the signing of the Louisiana Purchase, an Indian village, a buffalo hunt, a scene from the Cherokee "Trail of Tears," a cattle drive, and, last, a sod house. They finished on the last day of the fair, October 2, 1988.[31]

Sand sculptors Walter Schwartz and Paul Dawkins created a 300-ton, eight-foot tall by 260-foot long sculpture of Oklahoma history at the 1988 fair (Courtesy Daily Oklahoman*).*

Crowd-pleasing features such as the sand sculpture
had always been an important aspect of the fair's success.
Even Gristmill Jones and Henry Overholser knew that
entertainment, while not the sole purpose of the fair, was
an essential tool to get people to the fair where they were
offered the basic staples of education and boosterism.

Before World War II, staging entertainment to satisfy
that need had been relatively easy. Offer a flamboyant,
period piece stage show with a climax of fireworks and
you could fill the grandstand. By the 1980s that old
formula no longer worked. Whereas early fair visitors
had been primarily rural, the new audiences were largely
urban, with daily opportunities that ranged from movies
and theater to amusement parks and concerts. Hotz and
the fair board had to find a way to keep entertainment on
a high level, unique enough to draw big crowds, and
profitable enough to make money for the fair.

Hotz and the fair board faced the task of filling the
longer fair with headline entertainment every day. They
agreed that the State Fair Rodeo was a winner and

Edward C. Joullian, III, presents the
Teacher of the Year Award to Mary
Jane Bassett in 1993. State Superintendent of Education Sandy Garrett
(right) helps with the presentation.
(left) Jeff Cable provided the automobile for one year to the award winner.
(State Fair Archieves).

should be kept as the anchor of the lineup. With country music stars such as Merle Haggard and Loretta Lynn booked during each performance, good crowds and revenues were guaranteed. That left the other twelve nights open for something new.

One of the toughest decisions facing Hotz and the fair board was whether or not to keep the Ice Capades as the featured act. Since 1964 the costumed stage show on ice had been a fixture of each fair, but the format had become stale by modern standards, no longer able to attract sell-out crowds as it had in the past. Recognizing the fact that Baby Boomers were starting their own families, Hotz contracted instead with "Disney on Ice," which offered a variety of shows that appealed to young parents raised on the Mickey Mouse Club and the antics of Donald Duck, Goofy, and Chip and Dale.[32]

The remaining slot was filled by the Ringling Brothers Barnum & Bailey Circus, a three-ring entertainment spectacular that never seemed to age. Both additions made an immediate impact. Arena sales, stagnant at $521,403 in 1986, jumped to $1,128,686 in 1987 under the new format.[60] That formula was still working in 1993 when sales at the Arena hit an all-time high of $1,308,374.[33]

Hotz and his leadership team not only initiated changes such as a longer format and new entertainment packages, but also tackled the basic issue of the way the fairgrounds were operated, especially when it came to interim operations. Before 1983, the issue of interim operations had been one of conflict and disruption, a direct and distracting result of split management. When Hotz arrived on the scene, that perspective had already changed to one of maximizing a new tool for the traditional goals of the fair—education, entertainment, and boosterism.

Although interim operations had been part of the fair's history since 1907, the year-round use of fairgrounds facilities had increased dramatically after the big move in 1954. With plentiful open space in buildings designed for exhibitors, new clients lined up to become interim tenants.

The variety of events staged at the new fairgrounds was impressive. In 1958 the Last Frontier Council of the Boy Scouts rented space for the first annual Scout-O-Rama. More than 12,000 Boy Scouts, Cubs, and Explorers had attended sessions described as "heavy with the flavor of science and electronics." The next year, the same space was used for the Oklahoma City Home Show and Furniture Fair, previously held in the Zebra Room of the Municipal Auditorium. For eight days, the public was treated to sold out exhibits on home furnishings and a "$250,000 miracle gas kitchen." Another event moved from the Municipal Auditorium was the 7th Annual Southwest Boat Show which attracted more than 63,000 people during a seven-day run in April of 1960.[34]

Sporting events added to the list of interim users. One of the first attractions booked into the new Grandstand was stock car racing, which enjoyed unprecedented popularity in the 1950s. Local races were run every Friday night during the spring and summer, and circuit racing with big-name drivers took advantage of the 15,000-seat Grandstand facility. Racing became so popular that the Oklahoma City Junior Chamber of Commerce received permission from the city to build a drag strip south of the Grandstand.

The most visible long-term sports tenant was the 89ers baseball team, organized in 1961 as a AAA team of the Houston Astros, one of two major league expansion teams created that year. After voters of Oklahoma City approved a $465,000 bond issue to build a new baseball stadium on the fairgrounds, the 89ers took the field in 1963 and won the Pacific Coast pennant, the first league championship won by an Oklahoma City team since 1935.[35]

The Arena, after it was built in 1964, opened the fairgrounds to other sports. Ice hockey took the town by storm in 1965 when the Blazers, the farm team of the Boston Bruins, skated into the hearts of Oklahoma City

Flower competition remains a popular attraction for both competitors and visitors (State Fair Archives).

fans. Another action-packed event that sold out the 10,000-seat Arena was the National Finals Rodeo. The first NFR, also known as the world series of rodeo, had a seven-day run in December of 1965 and included the top fifteen cowboys in six events: saddle bronc, bareback, steer wrestling, calf roping, team roping, and the biggest crowd pleaser, bull riding. Few rodeo fans will ever forget competitors such as Larry Mahan, Dean Oliver, Jim Shoulders, and Oklahoma's own Freckles Brown, the aging cowboy who was the first man to ride the famous bull, Tornado.[36]

The distinction of being the very first event held in the Arena belonged to the All-Arabian Horse Show, booked into the fairgrounds for two days in September of 1965. More than 2,500 people attended each session as 121 registered Arabians competed in six events that included parade horse, jumper, cutting horse, and costume class. The last was the most colorful event of the show, as high-stepping horses went through their routines adorned with beaded and tasseled bridles and breast ornaments.[37]

Another major event held in the Arena that first year was the Oklahoma City Charity Horse Show, previously held at the Nichols Hills stables and the Stockyards Coliseum. In addition to classes for jumpers, hunters, harness teams, hackneys, and three- and five-gaited lady riders, the five-day event included a polo exhibition, a full orchestra, and singer Anita Bryant. Lavish parties were held all over town before, during, and after the horse show. Opening night drew a crowd of 4,000.[38]

By the time Hotz became president of the fair, the fairgrounds had already earned a national reputation as the premier site for horse shows. With the full support of Joullian and the fair board, he built on that tradition to make the facilities even better. Following a $5 million expansion program form 1990 to 1991, the fairgrounds could offer eight large barns with over 500,000 square feet of space, enough room to hold 3,000 horses, a 40,000-square-foot exercise arena, and the main Arena, connected to all barns. This physical plant, when added to the emphasis placed on recruiting and accommodating horsemen, made Oklahoma City the "Horse Show Capital of the World."

This international dominance was still evident in 1993 when fifteen horse shows were staged on the fairgrounds just during the fall fair. Included were the Morgan/Arabian Show, the Tennessee Walking Horse and Missouri Fox-Trotters Horse Show, the Dressage Show, the Team Penning Show, the Pony Pulling Contest, the Mule-Donkey Show, the Appaloosa Horse Show, the Draft Horse Overload Pull, the Draft, Harness and Driving Show, the Pinto Horse Show, the Paint Horse Show, the Working Cow Horse Show, the Quarter Horse Show, and the Buckskin Horse Show.

Rounding out the schedule in 1993 were 36 additional horse shows covering 104 event days during the interim months. National and international events included the Oklahoma Centennial, the National Appaloosa, the International Arabian Youth, the Championship Morgans, the huge International Quarter Horse Show, which was the largest horse show in the world, the Reining Futurity, and the Barrels Show. There also were three regional shows. In 1993 alone, the Oklahoma City Chamber of Commerce estimated that horse shows on the state fair-

The Oklahoma Publishing Company Pavilion added free country music shows as another attraction at the fair (Courtesy Daily Oklahoman*).*

grounds had an economic impact of $120 million.[39]

Horse shows, although the most visible and arguably the most important to Oklahoma City, were still just one segment of interim tenants of the fairgrounds. Altogether, the fairgrounds in 1993 was used for 1,600 event-days, a clear sign of the versatility built into the facilities. This growing importance of interim operations was not lost on the new chairman of the board, Edward Joullian, III. "There are cities which have newer facilities," he reported to the board of directors, "and to keep our events from going to those facilities, we have to increase service to the users." To Joullian, simply building show barns, stalls, and other facilities was not enough; the fair had to provide better ticketing, marketing, promotions, and food service.[40]

Together, Joullian, Hotz, and the fair board gradually crafted a new organizational structure with the flexibility of moving back and forth between running the fair and managing interim operations the other 348 days of the year. At the top of the organizational chart was the State Fair Division, which worked on traditional tasks associated with the seventeen-day fair, such as assembling the premium book, lining up judges, selling booth space, and coordinating general administration. Parallel to that was the Interim Division, a compartmentalized group of men and women who focused on purely interim events and everyday maintenance.[41]

To support both divisions, depending on the time of year and seasonal demands, Hotz assembled specialized teams. The Promotions Division was assigned the task of coordinating interim events such as stock car racing, boat shows, and horse shows. During the fair, that same staff coordinated all events in the Grandstand, Arena, and barns. Another innovation was the creation of the Food Service Division. It has become a profit center for Interim Events as well as the annual State Fair. Control of quality and price of food is an important aspect for all who come to the fairgrounds.

Another office that supported both the fair and interim events was the Ticketing Division. Prior to 1980 the fair staff had handled ticket sales only for the fair; interim users provided their own ticketing. Following the consolidation of interim management under the fair board, the fair contracted all ticketing to Ticket Master, a computerized service based at the University of Oklahoma.[42]

When Hotz was hired to run the State Fair of Oklahoma, he brought with him a thorough understanding of computerized ticketing and a staff member, Reba Jones, who could run the system. Within a year after losing Hotz and Jones, the University of Oklahoma offered the Ticket Master system to the fair board. In 1986 the board of directors authorized the purchase of the system, which included the rights under contract with owners of Ticket Master nationwide.

Armed with this new tool, the Ticketing Division expanded its services well beyond the fair and interim events. Jones and her staff provided services for all events in the Lloyd Noble Center at Norman and the Civic Center Music Hall and Myriad Convention Center in Oklahoma City. Other customers included the Lazy E Arena in Guthrie, OU football, Lyric Theater, and the Zoo Amphitheater at Lincoln Park. Other than making additional revenue for maintenance of the fairgrounds, the new system provided quality control for all events staged on the fairgrounds during the interim period. Just as importantly, the system was in place and working at peak efficiency when the fall fair rolled around each year.[43]

The benefits of the new organizational structure were apparent by 1989 when all records were broken during the seventeen-day fair. Contributing to that success were the weather and the centennial celebration of the land run of 1889. The weather was flawless throughout the fair, with mild temperatures, sunny days, and most importantly, no rain. The centennial, despite the disappointment of not featuring a world's fair, was commemorated with several events during the fair.

The opening ceremonies featured the rededication of the Tinker Air Force Base Memorial Aircraft Exhibit, which included a B-47 bomber, a B-52 bomber, a C-47 transport, and the prototype of the Aero Commander aircraft built at the Oklahoma City plant of Gulfstream Aircraft Company.

The closing ceremonies provided an appropriate finale as a capacity audience of 25,000 gathered in the Grandstand to witness a salute to history. With State Fair Chairman Edward C. Joullian III serving as master of ceremonies, the event started with the national anthem sung by noted vocalist Jody Miller and original musical scores depicting the land run of 1889 sung by the Men's Chorus of Oklahoma Christian University of Science and Arts. Governor Henry Bellmon offered a special message, followed by recognition of State Fair award winners.

G.T. Blankenship, chairman of the Centennial Commission, outlined the special events of the last year and recognized the efforts of participating cities throughout the Unassigned Lands. In a fitting ceremony, Frank Splendoria of the United States Department of the Interior presented title to the last unclaimed 160-acre parcel of land in the Unassigned Lands to the University of Oklahoma. Art Elbert, vice-president of the university, stated that the property would be used for research.

A time capsule also was dedicated at the closing ceremonies. In addition to the inclusion of a number of commemorative items, the capsule contained an official copy of President Benjamin Harrison's proclamation opening the Unassigned Lands by land run. The historical document from the National Archives was attached to a new proclamation signed by Secretary of the Interior Manuel Lujan.

The salute to history combined with good weather, improved grounds, and efficient management to make the fair of 1989 a record breaker. Attendance hit an all-time high of 1,815,260, while revenues topped out at $9,442,780. The numbers were a tribute to four years of fine-tuning programs and concentrating management efforts where they could make the greatest impact.[44]

The momentum generated by the fair in 1989 continued into the new decade. In 1993, reaching back to one of the basic building blocks of the fair, the board chose "Excellence in Education" as the theme. One entire pavilion was filled with educational exhibits sponsored by universities, technical training schools, computer companies, and state agencies. And once again the fair sponsored the "Teacher of the Year Award," with a special presentation made in front of 15,000 spectators at the closing ceremonies.

The fair also offered a unique combination of things old and new. Rod Link again provided a wide variety of thrill rides and games but with a new emphasis on cour-

The Midway, with more than 90 rides, remains a major source of revenue for the fair, making free attractions possible at the fair (Courtesy Daily Oklahoman).

tesy and appearance among ride attendants. Independent contractors added to the appeal, especially with a wild tandem seat ride that was a variation on bunjee jumping and a vertical wind tunnel that simulated the feeling of free-fall sky diving. Large crowds lined up not only to ride the rides, but just to watch the show.[45]

Convinced that the traditional fall fair was expanding and improving and pleased with the quality and quantity of interim events, the fair board turned its attention to yet another new opportunity, a spring fair. On August 19, 1992, Chairman Joullian appointed a special committee to look into the possibility of such a fair. Appointed to the committee were Clay Bennett, chairman, and members Boots Hall of Fred Jones Industries, Rusty Noble of SEA Cattle Company in Ardmore, and Bill Pirtle of Oklahoma Natural Gas Company.[46]

The primary purpose of the spring fair was to strengthen the 4-H and FFA Junior Livestock Show, which had become the largest of its kind in the nation. For years it had been sponsored by the Oklahoma City Chamber of Commerce and its offspring, the Sirloin Club of Oklahoma and the Southwest American Livestock Foundation. The Sirloin Club was unique because its members supported Farm Youth Livestock Shows throughout the state by raising funds to purchase livestock. Despite the long-term successes of these groups, additional support was needed as the Chamber de-em-

Clowns remain a feature of the Ringling Brothers and Barnum and Bailey Circus, which plays the fair each year. The State Fair of Oklahoma is the only site where Ringling Brothers book both the circus and Disney on Ice back-to-back (Courtesy Ringling Brothers and Barnum and Bailey Circus).

phasized farm youth as an aspect of its program.

The special committee, chaired by Bennett, held a series of meetings with all parties involved to forge a new partnership between the State Fair of Oklahoma, the Southwest American Livestock Foundation (SWALF), and the Oklahoma 4-H & FFA Junior Livestock Show. Under the new arrangement, the State Fair would be responsible for providing facilities and management services, including the collection of entry fees, other revenues, and the disbursement of premiums. The State Fair and SWALF would constitute the governing body for the event. The Oklahoma City Chamber of Commerce and the Sirloin Club of Oklahoma agreed to continue leadership and fund-raising support.[47]

Officially called the "Agriculture and Youth Expo," the spring fair was scheduled for March 25 to April 2, 1994, the first time in the history of Oklahoma that such an event would be attempted. At the announcement, Joullian emphasized that "the agriculture industry is essential to securing a bright future for the State of Oklahoma. The more Oklahomans know about the agriculture industry the greater they will appreciate it."[48]

The confidence to stage a spring fair was a direct result of the goals set and achieved by the fair board,

Hotz, and his staff. Their ability to take on another major exposition, in addition to the ever-increasing pace of interim operations and an expanded fall fair, showed that everyone associated with the fair—the city council, the people of Oklahoma City, and the vendors who make each fair such a success—were convinced that the future of the fair was in good hands.

That success set the stage for a major leap forward when Mayor Ronald Norick proposed a major new capital program to expand and improve city facilities. Dubbed

A parachutist, part of an Army demonstration team, descended on the fairgrounds in 1992 (State Fair Archives).

"MAPS," short for Metropolitan Area Projects, the initiative included major projects such as a new sports arena, a new baseball stadium, renovations to Civic Center Music Hall, a riverfront canal through Bricktown, a new Library, and fairgrounds improvements. As the mayor and several committees assembled a needs list, they asked the fair board for a list of potential projects.

In a letter to Mayor Norick, Hotz and Joullian asked for $16,989,000 to "update existing fairgrounds' facilities...to bring these facilities up to building code and Americans with Disabilities Act guidelines...and to complete all phases of improvements to the livestock and horse show facilities." In their opinion, such investment would "attract new business, but more importantly, retain those events which we now have." The message was clear. If the State Fair of Oklahoma was to remain the "Horse Show Capital of the World," major improvements had to be made.[49]

Some of the suggested projects were essential if not dramatic. More than $5 million was needed to replace electrical systems, plumbing, water and sewer lines, heating and air conditioning, and insulation in the Kitchens of America Building, the International Trade Center, the Made in Oklahoma Building, the Hobby, Arts, and

Crafts Building, the Modern Living Building, the Carriage Hall, the Travel and Transportation Building, and barns 3, 4, 5, 7, 8, and 9. This series of improvements included covering the central courtyard between the Travel and Transportation Building and the Carriage Hall.

Another $5 million was needed for improvements to the grounds, especially electrical systems, water and sewer lines, street surfacing, street lights, curbs, and gutters. More than $1 million would surface both the north and south parking lots, while $1.5 million would build new drainage ditches and channels from the fairgrounds to Reno Street and May Avenue.

More visible to the public would be another $5 million needed just in the Arena. The list of projects included work on heating and air conditioning, utility extension, installation of acoustical surfaces on walls and ceiling, and construction of new ticket booths, lobby, restrooms, and concession stands. The most dramatic improvement would be a new 40,000-square-foot exhibit hall attached to the east side of the Arena. Such an expansion would

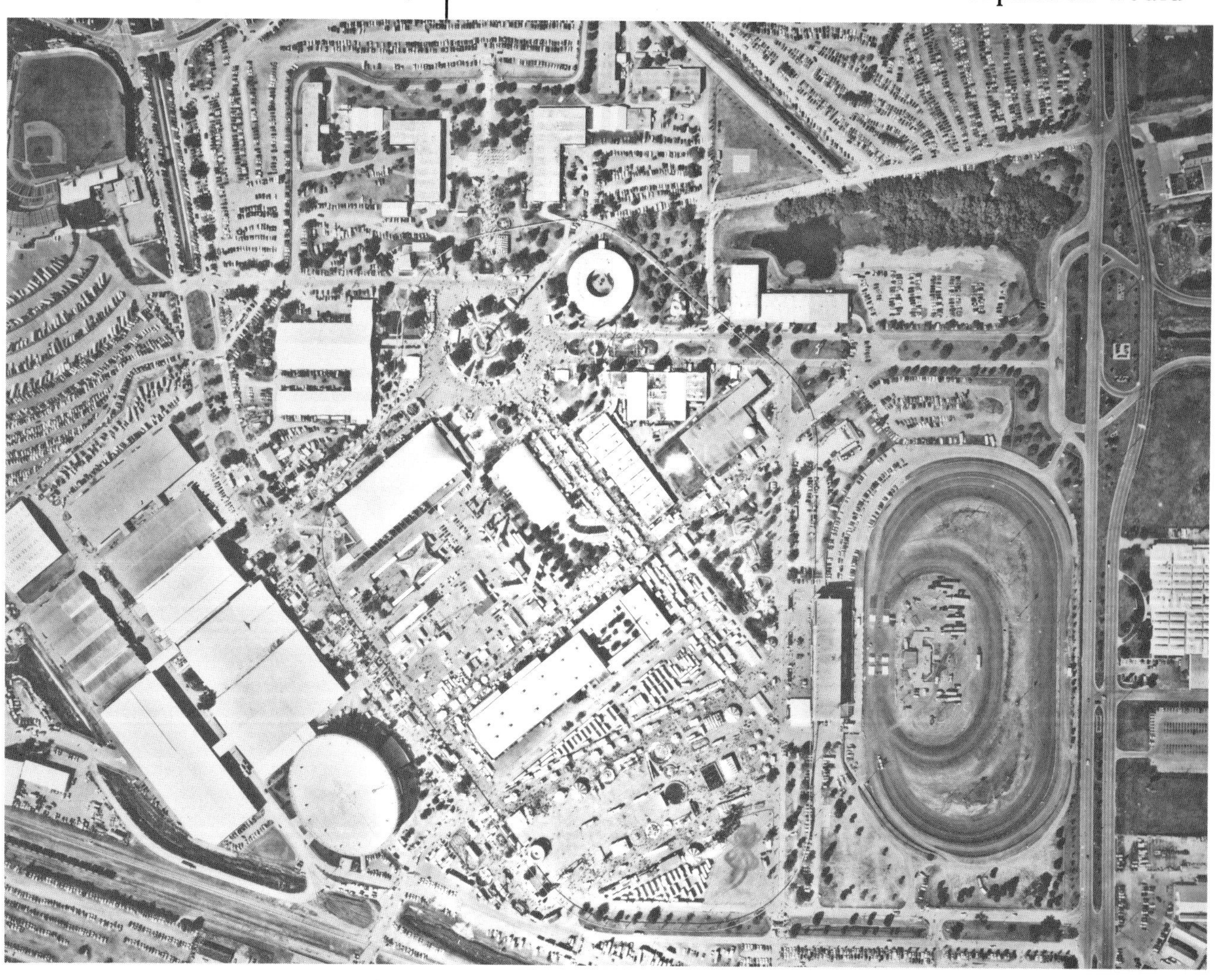

This aerial view of the fairgrounds was taken in 1993 (State Fair Archives).

The State Fair of Oklahoma, in 1994 as in 1907, plays an important role in the life of Oklahoma City, the State, and the nation (State Fair Archives).

greatly enhance the facility as a site for conventions and trade shows.

The highest profile, and the greatest potential impact on the important horse show business, was the plan for a world class sales facility with seating for 2,500 and special seating for 700 in a VIP and buyers' area. With that project included, the total wish list was $16,989,000, the single largest capital plan in the history of the fair.[50]

The mayor and city council included the $16 million request from the fair board in a proposed sales tax issue that totaled $238,388,000. If approved by the voters of Oklahoma City, a one-penny sales tax would be levied for five years with the earmarked income dedicated to the capital projects. After a well-orchestrated promotional campaign asked voters whether or not they wanted Oklahoma City to be a big-league city, the polls were opened and the ballots were counted. On December 14, 1993, the special tax for "MAPS" was approved. The people of Oklahoma City would provide the money needed to keep the State Fair of Oklahoma one of the top five expositions in the country.[51]

At the Board of Directors meeting on February 16, 1994, Chairman of the Board Edward Joullian expressed his faith in the future. "This bond issue is not only a significant turning point for the history of Oklahoma City—it is a turning point in the history of the State Fair of Oklahoma. With these funds we will continue to serve the people of Oklahoma. We will continue to build on the fair's heritage of education, entertainment, and economic development." If history is any indication, that prediction will come true.

DIRECTORS OF THE STATE FAIR OF OKLAHOMA, 1907-1994

Classen, A. H., Board Member,	1907*
Gloyd, S. M., Board Member,	1907*
Land, L. L., Board Member,	1907*
Putnam, I. N., Board Member,	1907*
Frances, J. L., Board Member,	1907*
Shelley, F. H., Board Member,	1907*
Bath, V. E., Board Member,	1907-1909*
Flick, Mattie H., Board Member,	1907-1909*
Jones, C. G., Board Member,	1907-1909*
Jordan, F. M., Board Member,	1907-1909*
Keller, C. H., Board Member,	1907-1909*
Sites, C. P., Board Member,	1907-1909*
Heyman, S. G, Board Member,	1907-1912*
Overholser, Henry, Board Member,	1907-1915*
Wilkin, J. L., Board Member,	1907-1916*
Bass, J. M., Board Member,	1908-1909*
Longfellow, G. N., Board Member,	1908-1909*
Mahan, Isaac S., Board Member,	1908-1923*
Huckins, Joseph, Jr., Board Member,	1908-1927*
McQuilken, Vera G., Board Member,	1908-1957*
Atwood, Weston, Board Member,	1910-1913*
Fields, John, Board Member,	1910-1914*
Brock, S. L., Board Member,	1910-1917*
Stone, G. B., Board Member,	1910-1917*
Ashton, Orin, Board Member,	1910-1923*
Colcord, C. F., Board Member,	1912-1917*
Mee, William, Board Member,	1913*
Owen, J. M., Board Member,	1913-1950*
Noble, John N., Board Member,	1915-1925*
Warren, J. F., Board Member,	1915-1925*
Everest, J. H., Board Member,	1916-1932*
Kerr, George G., Board Member,	1918*
Gustin, A. M., Board Member,	1918-1923*
Myer, Joseph, Board Member,	1919-1925*
O'Neil, John E., Board Member,	1919-1928*
Hemphill, Ralph, Board Member,	1919-1953*
Hill, John L., Board Member,	1924-1932*
Lamb, Floyd S., Board Member,	1924-1951*
Campbell, A. O., Board Member,	1925-1934*
Sohlberg, George G., Board Member,	1926-1927*
Huntley, H. W., Board Member,	1926-1931*
Monroney, A. E., Board Member,	1928*
Nichols, G. A., Board Member,	1928-1934*
Dietz, Oscar, Board Member,	1928-1951*
Bulkley, W. S., Board Member,	1929-1930*
Owens, J. F., Board Member,	1929-1934*
Flanagan, George, Board Member,	1931*
Baker, John R., Board Member,	1932-1942*
Day, C. C., Board Member,	1933-1939*
Benzel, R. J., Board Member,	1933-1943*
Browne, Virgil, Board Member,	1933-1980*
Frederickson, George, Board Member,	1935-1943*
Davis, George A., Board Member,	1935-1947*
Rainey, Judge Robert, Board Member,	1936-1971*
Martin, J. Frank, Board Member,	1940-1968*
Callahan, L. D., Board Member,	1944-1946*
Dee, Tom J., Board Member,	1944-1957*
Earp, Ancel, Board Member,	1944-1976*
Macklanburg, Louis A., Board Member,	1948-1965*
Peebly, Robert L., Board Member,	1949-1960*
Owen, Nelson, Board Member,	1951-1972*
Anthony, C. R., Board Member,	1952-1976*
Kennedy, Donald S., Board Member,	1952-present
Taylor, Ross, Ex-Officio, City Manager,	1953-1954*
Gill, Wm. Sr., Board Member,	1955-1970*
Dulaney, Luther T., Board Member,	1955-1973*
Gill, Wm. Jr., Ex-Officio, City Manager,	1955-1956;
Board Member	1963-1976*
James, Dan W., Board Member,	1955-1988*
McGee, Dean A., Board Member,	1955-1989*
Gaylord, Edward L., Board Member,	1955-present
Draper, Stanley C., Sr., Ex-Officio,	1957;
Board Member	1958-1976*
Stirling, Sheldon, Ex-Officio,	1951-1961
Groh, H. B., Board Member,	1958-1959
Wolf, Larry, Board Member,	1951-1975*
Holderby, Oscar, Board Member,	1958-1976*
Sturm, George N., Board Member,	1958-1983*
Smith, Dr. Kenneth E., Board Member,	1958-present
Wileman, Ben C., Board Member,	1958-present
Kirkpatrick, John E., Board Member,	1959-present
Kilpatrick, John, Jr., Board Member,	1959-present
Roederer, E. L. "Jim", Board Member,	1960-1980*

Rucks, Joseph G., Board Member, 1960-1988*
Doherty, James J. Jr., Board Member, 1961-1963
Mundell, Maj. Gen. Lewis,
    Board Member, 1961-1963
Bayne, Lewis N., Board Member, 1961-1964
Smith, Edward C., Board Member, 1961-1965
Saunders, Orval O. "Sandy",
    Board Member, 1961-1993*
Hoghn, Harry H., Board Member, 1962
Holton, F. W., Advisory, 1962
Luttrell, Robert T., Ex-Officio,
    City Manager, 1962-1963*
Needham, I. H., Advisory, 1962-1964
Parker, Dr. Jack F., Ex-Officio, 1962-1966
Cooper, Hal, Advisory, 1964-1970*
Boecher, Roy, Advisory, 1962-1974*
Camp, John N. "Happy,"
    Board Member, 1962-1987*
Bellmon, Gov. Henry, Ex-Officio, 1963-1966;
    Ex-Officio, 1987-1990;
    Advisory, 1991-present
Mosier, O. M. "Red", Board Member, 1963-1966*
Finney, W. D. "Jim", Advisory, 1963-1985
Tintsman, Robert M., Ex-Officio, 1964-1967
Hutton, Harold, Advisory, 1964-1968
McNickle, Maj. Gen. M. F.,
    Board Member, 1964-1968*
Massad, Maj. Gen. Ret. E. L. "Mike",
    Board Member, 1964-1992*
Allen, Lloyd K., Advisory, 1965
Shirk, George, Ex-Officio, 1965-1966*
Webb, H. C., Board Member, 1965-1968*
Macklanburg, Robert Jr.,
    Board Member, 1965-1974*
Gosselin, E. L. "Les", Board Member, 1965-1977*
Vose, C. A., Sr., Board Member, 1965-1984*
Bolen, Ralph L., Board Member, 1965-present
Hunt, Joe H., Board Member, 1966
Horton, W. G. "Bus", Board Member, 1966-1977*
Norick, Mayor James H., Ex-Officio, 1967-1970;
    Board Member, 1990-present
Bartlett, Gov. Dewey, Ex-Officio, 1967-1970*
Lillard, Dr. Bill, Ex-Officio, 1967-1975
Parker, Wayne, Board Member, 1967-1984
Parsons, John R., Board Member, 1967-present
Oldland, Robert H.,
    Ex-Officio, City Manager, 1968-1970
Johnson, Maj. Gen. George M.,
    Board Member, 1968-1971
Strasbaugh, Paul B., Board Member, 1968-present

Young, Stanton L., Board Member, 1968-present
Tyree, James E., Board Member, 1969-1972
Lyons, Patrick B., Board Member, 1970-1971
Healey, Burke, Advisory, 1970-1981
Hitch, H. C. "Ladd", Jr., Advisory, 1970-1986
Edwards, Roy V., Board Member, 1970-present
Pittinger, James, Board Member, 1971-1973
Ross, Nate, Ex-Officio, City Manager, 1971-1973
Hall, Gov. David, Ex-Officio, 1971-1974
Latting, Mayor Patience, Ex-Officio, 1971-1982
Hulsey, William W., Board Member, 1971-1982*
Reinbold, Maj. Gen. Richard D.,
    Board Member, 1972
Knotts, Max, Board Member, 1972-present
Smith, Maj. Gen. W.Y.,
    Board Member, 1972-1973
Borum, Maj. Gen. F. S.,
    Board Member, 1972; Advisory, 1973-1978*
Sims, Dr. William E., Advisory, 1973
Painter, Pat, Ex-Officio, 1973-1974
Randolph, Maj. Gen. James G.,
    Board Member, 1973-1976
Grubin, S.M., Board Member, 1973-1980
Anthony, Ray, Board Member, 1973-present
Breeding, M. O. "Bud,"
    Board Member, 1973-present
Dulaney, L. Thomas, Jr.,
    Board Member, 1973-present
McMahan, Howard D., Ex-Officio,
    City Manager, 1974-1976
Harlow, James G., Jr., Board Member, 1974-present
Kamm, Dr. Robert B., Board Member, 1975-present
Boren, Gov. David, Ex-Officio, 1975-1979
Springer, Jack, Advisory, 1975-1983
Griggy, Kenneth J., Board Member, 1975-present
English, Dr. Thomas, Ex-Officio, 1976-1977
Schneider, Maj. Gen. Carl G.,
    Board Member, 1976-1977
Smith, Dr. Thomas J., Ex-Officio, 1976-1978
Cook, James J., Ex-Officio,
    City Manager, 1976-1981
Jones, Orville A., Board Member, 1976-1981
McEldowney, James R., Board Member 1976-1989*
Swisher, Bill, Board Member, 1976-present
Boger, Dr. Larry J., Board Member, 1977-present
Fox, Maj. Gen. Cecil E., Board Member, 1977-1980
Braun, Edward J., Board Member, 1977-present
Harrison, Richard D., Board Member, 1977-present
Peters, James R., Board Member, 1978-1981*
Payzant, Dr. Thomas W., Ex-Officio, 1978-1983

Stewart, James E., Board Member, 1978-present
Nigh, Gov. George, Ex-Officio, 1979-1984
Banowsky, William S., Board Member, 1979-1985
Edwards, Maj. Gen. Jay T.,
    Board Member 1980-present
Johnson, Scott, Ex-Officio,
    City Manager, 1981-1984
Cook, Edward H., Ex-Officio, 1981-1987
Hogan, Dan, III, Board Member, 1981-present
Light, Maj. Gen. James E. Jr., Board Member, 1982
Wright, Dr. Donald L., Ex-Officio, 1982-1984
Gaugler, Richard L., Board Member, 1982-present
Joullian, Edward C., III,
    Board Member, 1982-present
Marshall, Gerald R., Advisory, 1983-1984
Basey, Albert L., Board Member, 1983-1985
Burpee, Maj. Gen. Richard A.,
    Board Member, 1983-1986
Coats, Andy, Ex-Officio, 1983-1987
Bollenbach, Irvin K., Board Member, 1984-present
Browne, Robert F., Board Member, 1984-present
Carey, William V., Jr., Board Member, 1984-present
Hall, Fred J., Board Member, 1985-1986
Smith, Lee Allan, Advisory, 1985-1987;
    Board Member, 1988-present
Horton, Dr. Frank, Board Member, 1985-1988
Hotz, Donald J., Board Member, 1985-present
Bowden, Maj. Gen. William P.,
    Board Member, 1986-1987
Bennett, Clayton I., Board Member, 1986-present
Gannon, Peter A., Board Member, 1986-present
Pirtle, William N., Board Member, 1986-present
Childers, Terry L., Ex-Officio,
    City Manager, 1987-1988
Leonard, Tim, Advisory, 1987-1991
Moore, R. W. "Dick", Board Member, 1987-1991*

Steller, Arthur W., Ex-Officio, 1987-1992
Bishop, Bill, Board Member, 1987-1993*
Beutler, Lynn, Advisory, 1987-present
Hall, Brooks Jr., Board Member, 1987-present
McPherson, Frank A., Board Member, 1987-present
Norick, Mayor Ronald W., Ex-Officio, 1987-present
Ingle, Clyde, Advisory, 1988-1989
Caldwell, Royce, Board Member, 1988-1990
Lyles, Lloyd J. "Jimmy", Ex-Officio, 1988-1992
Campbell, Dr. John R., Advisory, 1988-1993
McClure, Dr. Homer C., Board Member, 1988-present
Hearn, Paula, Ex-Officio, City Manager, 1989-1990
Stone, Wayne D., Board Member, 1989-1992
Martin, Edmund O., Board Member, 1989-present
Mashburn, J. W., Board Member, 1989-present
Spiers, Maj. Gen. Joseph K.,
    Board Member, 1989-1994
Ellis, J. B., Board Member, 1990-present
Durrett, William, Board Member, 1990-present
Bown, Donald D., Ex-Officio,
    City Manager, 1990-present
VanHorn, Dr. Richard L., Advisory, 1990-1994
Ackerman, Ray, Ex-Officio, 1991
Buchanan, F. G. "Buck," Ex-Officio, 1991-present
Noble, Russell "Rusty", Board Member, 1991-present
Townsend, Kenneth W., Board Member, 1991-present
Walters, Gov. David, Ex-Officio, 1991-present
Werries, E. Dean, Board Member, 1991-present
Mathis, Bill, Board Member, 1992-present
Moore, Bob, Board Member, 1992-present
Garrett, Sandy, Ex-Officio, 1992-present
Gaylord, Edward King, II, Advisory, 1992-present
Perry, Russell M., Board Member, 1992-present
Van Rysselberge, Charles H.,
    Ex-Officio, 1993-present
Smith, John M., Board Member, 1994-present

*DECEASED

NOTES

Chapter One:

[1]Muriel H. Wright, "The Indian International Fair at Muskogee," *The Chronicles of Oklahoma*, LXIX (Spring, 1971), 14-50.

[2]The Oklahoma Star (McAlester, Choctaw Nation), November 6, 1874; *The Vindicator* (New Boggy, Choctaw Nation), May 8, 1875.

[3]*Ibid.*; "Resolution of the General Council of the Indian Territory, at Okmulgee, May 7th, 1875," as printed in *The Chronicles of Oklahoma*, LXIX (Spring, 1971), 50; Interview with Will R. Robison, Indian-Pioneer Papers, VIII, 532-540, Oklahoma Historical Society.

[4]Wright, "The Indian International Fair at Muskogee," 14-17; "Premium List, Constitution, Acts, and Rules of the Fifteenth Exhibit of the Indian International Agricultural Society and Fair Association, September 17, 18, 19, 20, 1892, at Muskogee, Indian Territory," a pamphlet in the collections of the Oklahoma Historical Society.

[5]*Oklahoma Gazette* (Oklahoma City), September 17, 1889.

[6]*Ibid.*

[7]*Ibid.*

[8]*Oklahoma Gazette*, April 23, 1890.

[9]*Oklahoma Gazette*, January 21, 1891.

[10]*Oklahoma Gazette*, February 19, 1892; February 24, 1892.

[11]"Charles Gasham Jones," a biographical sketch prepared through the Writers' Project of the Works Progress Administration, July 7, 1937, in the files of the Oklahoma Historical Society.

[12]*Oklahoma Gazette*, February 24, 1892.

[13]*Oklahoma Gazette*, March 12, 1892.

[14]*Oklahoma Gazette*, April 4, 1892.

[15]*Oklahoma Gazette*, April 5, 1892, April 6, 1892.

[16]*Oklahoma Gazette*, April 11, 1892.

[17]*Ibid.*

[18]*Oklahoma Gazette*, June 29, 1892, July 9, 1892, July 13, 1892.

[19]"Premium List of the First Annual Fair of the Oklahoma Territorial Fair Association, October 4th to 8th, 1892," a document in the files of the State Fair of Oklahoma, Oklahoma City. This booklet also includes the Constitution and Bylaws of the organization.

[20]"James Geary," a biographical sketch in *Portrait and Biographical Record* (Chicago: Chapman Publishing Company, 1901), 31-32.

[21]W.F. Kerr, *The Story of Oklahoma City, Oklahoma: The Biggest Little City in the World* (Chicago: The S.J. Clarke Publishing Company, 1922), 87-88; Roy Stewart, *Born Grown: An Oklahoma City History* (Oklahoma City: Fidelity Bank, 1974), 7, 20, 21; Patricia Lester, "William J.

McClure and the McClure Ranch," *The Chronicles of Oklahoma*, LVIII (Fall, 1980), 296-307.

[22]*Oklahoma Gazette*, August 27, 1892.

[23]"Premium List of the First Annual Fair, 1892," 11-13.

[24]*Ibid.*, 23-101.

[25]*Ibid.*, 103; *Oklahoma Evening Gazette*, October 3, 1892.

[26]*Oklahoma Evening Gazette*, October 5, 1892.

[27]*Oklahoma Evening Gazette*, October 6, 1892, October 7, 1892.

[28]*Oklahoma Daily Times Journal* (Oklahoma City), October 7, 1892.

[29]Bob L. Blackburn, *Heart of the Promised Land: An Illustrated History of Oklahoma County* (Los Angeles: Windsor Publishing Company, 1982), 52-82.

[30]*Ibid.*

[31]*Oklahoma Press-Gazette* (Oklahoma City), October 26, 1893.

[32]"Premium Book for the 1894 Oklahoma Territorial Fair," a typescript in the files of the State Fair of Oklahoma; *Daily Oklahoman* (Oklahoma City), September 20, 1894, September 25, 1894; *Oklahoma Times Journal* (Oklahoma City), September 28, 1894.

[33]*Daily Oklahoman*, October 5, 1896, September 26, 1897, Septemer 30, 1897.

[34]*Corporation Records*, Territory of Oklahoma, Volume 5, August 27, 1897, 105-106, in the Archives Division of the Oklahoma Historical Society; *The Stillwater Gazette* (Stillwater, Oklahoma), September 16, 1897.

[35]Blackburn, *Heart of the Promised Land*, 82-97.

[36]*Ibid.*

[37]*Ibid.*

[38]"A Territorial Triumph: The Street Fair at Oklahoma City, October 10th to 15th, 1898," *McMasters' Magazine*, X (November, 1898), 4-9.

[39]*Ibid.*, 9-13.

[40]*Ibid.*

[41]*Daily Oklahoman*, September 15, 1899, September 19, 1899, September 20, 1899, September 21, 1899, September 22, 1899, September 23, 1899, September 24, 1899.

[42]Charles Colcord, *The Autobiography of Charles F. Colcord, 1859-1934* (Tulsa: C.C. Helmerich, 1970).

[43]*Daily Oklahoman*, September 7, 1902.

[44]*Daily Oklahoman*, September 10, 1902, September 14, 1902.

Chapter Two:

[1]Bureau of the Census, *Thirteenth Census of the United States, taken in the Year 1910, Statistics of Oklahoma* (Washington: Government Printing Office, 1914)., 621-622.

[2]*Ibid.*, 631.

[3]*Ibid.*; Bureau of the Census, *Statistical Abstract of the*

United States, 1907 (Washington: Government Printing Office, 1908), 565-567.

[4]A.W. McKeand, "Why, What and Why," *Sturm's Oklahoma Magazine*, Vol. 3 (November, 1906), 14.

[5]*Ibid.*, 11-12; *Census of 1910*, 679.

[6]McKeand, "Why, What and Why," 11.

[7]*Ibid.*, 11-14.

[8]Bureau of the Census, *Bulletin 89: Population of Oklahoma and Indian Territories, 1907* (Washington: Government Printing Office, 1907), 7; *Census of 1910*, 569, 584.

[9]Joseph B. Thoburn, "Product of Pulling Together," *Sturm's Oklahoma Magazine* (February, 1910), 22-25.

[10]*Ibid.*, 26.

[11]*Ibid.*, 26-27.

[12]*Ibid.*

[13]Frederick S. Barde, "What a State Fair Means to Oklahoma," *Sturm's Oklahoma Magazine*, Vol. 6, no. 4, 22-23.

[14]*Ibid.*, 24.

[15]W.M. Sturgis, "Oklahoma's First State Fair," *Sturm's Oklahoma Magazine*, Vol. 5, no. 1, 15.

[16]"Minutes of the State Fair of Oklahoma Board of Directors," January 18, 1907, Archives of the State Fair of Oklahoma, Oklahoma City, Oklahoma. All subsequent references to minutes of the Board of Directors will be written as Board Minutes.

[17]Board Minutes, January 18, 1907.

[18]*Ibid.*

[19]*Corporation Records*, Territory of Oklahoma, January 21, 1907; Board Minutes, January 25, 1907.

[20]Colcord, *The Autobiography of Charles F. Colcord*; James T. Grady, ed., "I.M. Putnam," in *The State of Oklahoma: Its Men and Institutions* (Oklahoma City: The Daily Oklahoman, 1908), 37.

[21]Grady, ed., "Vincent L. Bath," in *The State of Oklahoma: Its Men and Institutions*, 77; "Francis M. Jordan, M.D.," in *Portrait and Biographical Record of Oklahoma* (Chicago: Chapman Publishing Co., 1901), 679-680; Board Minutes, February 12, 1907.

[22]Board Minutes, February 25, 1907.

[23]Clay Bailey, "State Fair of Texas," in Walter Prescott Webb, ed., *The Handbook of Texas* (Austin: The Texas State Historical Association, 1952), vol. 2, 662.

[24]Board Minutes, March 19, 1907, March 29, 1907, April 1, 1907.

[25]Board Minutes, April 12, 1907.

[26]Board Minutes, April 25, 1907.

[27]Board Minutes, May 10, 1907, May 15, 1907.

[28]Board Minutes, June 6, 1907.

[29]Board Minutes, June 11, 1907, June 17, 1907, June 27, 1907, June 24, 1907

[30]"The Oklahoma State Fair," *Sturm's Oklahoma Magazine*, Vol. 5 (October, 1907), 10.

[31]*Ibid.*

[32]*Ibid.*, 10-12.

[33]*Daily Oklahoman*, October 6, 1907.

[34]*Ibid.*

[35]"The Oklahoma State Fair," *Sturm's Oklahoma Magazine*, Vol. 5 (October, 1907), 10.

[36]*Daily Oklahoman*, October 10, 1907.

[37]*Daily Oklahoman*, October 12, 1907.

[38]*Daily Oklahoman*, October 10, 1907.

[39]*Daily Oklahoman*, October 11, 1907.

[40]*Ibid.*

[41]*Daily Oklahoman*, October 16, 1907.

[42]*Daily Oklahoman*, October 12, 1907.

[43]*Daily Oklahoman*, October 15, 1907.

[44]*Daily Oklahoman*, October 8, 1907.

[45]*Daily Oklahoman*, October 13, 1907.

[46]*Ibid.*

[47]*Ibid.*

[48]Diane B. Haser-Harris, "Horse Racing in Early Oklahoma," *The Chronicles of Oklahoma*, Vol. 64 (Spring, 1986), 5-6.

[49]*Ibid.*, 7-9.

[50]*Ibid.*, 10-11.

[51]Board Minutes, January 18, 1907.

[52]*Daily Oklahoman*, October 11, 1907.

[53]*Ibid.*

[54]*Daily Oklahoman*, October 13, 1907.

[55]*Ibid.*

[56]*Daily Oklahoman*, October 16, 1907.

Chapter Three:

[1]Board Minutes, October 21, 1907.

[2]Ralph Jones, "Henry Overholser and the Overholser Mansion," an unpublished manuscript in the files of the Oklahoma Historical Society, Historic Sites Division.

[3]Marion Rock, *Land of the Fair God* (Oklahoma City: np, 1900).

[4]Jones, "Henry Overholser and the Overholser Mansion."

[5]Board Minutes, November 12, 1907, November 14, 1907.

[6]Board Minutes, December 24, 1907.

[7]Joseph Thoburn and Muriel Wright, "Isaac Shepherd Mahan," in *Oklahoma: A History of the State and Its People* (New York: Lewis Historical Publishing Company, 1929), vol. 4, 476-77.

[8]Undated newspaper clipping in "Scrapbook, 1908-09," in the Archives of the State Fair of Oklahoma. Hereafter, all references to undated and untitled newspaper clippings in the Archives of the State Fair of Oklahoma will be referred to as "Scrapbook," with the year; when titles or newspapers are identified in the files, the information will be included in the citation.

[9]"Puts a Fortune in the Fair," Scrapbook, 1908-09.

[10]Board Minutes, June 6, 1908.

[11]Board Minutes, July 14, 1908; Scrapbook, 1908-09.

[12]Scrapbook, 1908-09; Board Minutes, April 14, 1908, July 14, 1908.

[13]"Plans Laid for Third State Fair," Scrapbook, 1908-09.

[14]Blackburn, *Heart of the Promised Land: An Illustrated History of Oklahoma County*, 82-98.

[15]*Ibid.*

[16]Board Minutes, May 25, 1909; Scrapbook, 1908-09; "Canal Tour in New Fair Show," Scrapbook, 1908-09.

[17]Board Minutes, May 25, 1909; Scrapbook, 1908-09.

[18]Board Minutes, January 3, 1910, April 13, 1910, May 27, 1910, February 7, 1911.

[19]Board Minutes, April 13, 1910; "Pavilion Under Way," Scrapbook, 1910.

[20]Blackburn, *Heart of the Promised Land: An Illustrated History of Oklahoma County*, 102-103.

[21]*Ibid.*, 104.

[22]"Juvenile Agricultural Organization," Scrapbook, 1910.

[23]"Aggie Class for State Fair Dates," Scrapbook, 1910.

[24]*Ibid.*

[25]"Every County to be Represented," *Daily Oklahoman*, July 9, 1911.

[26]"Governor Condemns Exodus from Farm; Advises Rural Life," *Oklahoma City Times*, September 27, 1916.

[27]"The Fair is a School for all the Folks," *The Oklahoma Farmer Stockman*, August 25, 1916.

[28]*Oklahoma Farmer Stockman*, October 17, 1912.

[29]*Oklahoma City Times*, July 22, 1913.

[30]*Daily Oklahoman*, October 2, 1916.

[31]"Dairy Exhibit Newest Feature," *Daily Oklahoman*, August 6, 1916.

[32]"Oklahoma City to Have Horse Show," Scrapbook, 1910.

[33]*Thomas Tribune* (Thomas, Oklahoma), August 24, 1911; Oklahoma

[34]Scrapbook, 1910.

[35]Scrapbook, 1911.

[36]*Daily Oklahoman*, April 17, 1913.

[37]*Oklahoma City Times*, June 7, 1912; Board Minutes, November 20, 1911, April 18, 1912.

[38]*Oklahoma City Times*, September 22, 1913; Board Minutes, September 24, 1913.

[39]Board Minutes, August 1, 1913.

[40]*Daily Oklahoman*, August 10, 1913.

[41]*Collinsville Times* (Collinsville, Oklahoma), August 19, 1913.

[42]*Daily Oklahoman*, July 27, 1916, September 26, 1916.

[43]*Daily Oklahoman*, September 28, 1916.

[44]Scrapbook, 1913.

[45]Scrapbook, 1910.

[46]*Daily Oklahoman*, February 1, 1911.

[47]*The Democrat-Record* (Idabel, Oklahoma), June 27, 1911.

[48]*Daily Oklahoman*, February 13, 1917.

[49]*Chickasha Daily Express*, February 12, 1917; *Bartlesville Examiner*, February 16, 1917.

[50]*Muskogee Democrat*, February 15, 1917.

[51]Board Minutes, November 11, 1914.

[52]Board Minutes, September 8, 1915; Scrapbook, 1913.

[53]*Daily Oklahoman*, August 22, 1916.

Chapter Four:

[1]*Oklahoma City Times*, February 26, 1917; *Daily Oklahoman*, April 25, 1917; *Tulsa Democrat*, May 1, 1917; *Daily Oklahoman*, May 8, 1917.

[2]Nancy Wiley, *The Great State Fair of Texas: An Illustrated History* (Dallas: Taylor Publishing Company, 1985), 48-50.

[3]*Ibid.*, 51-52.

[4]Board Minutes, November 24, 1916, February 5, 1917, February 28, 1917.

[5]Board Minutes, January 14, 1918.

[6]*Ibid.*

[7]*Ibid.*

[8]Scrapbook, 1918.

[9]*Ibid.*

[10]*Oklahoma City Times*, April 26, 1917; *Oklahoma News*, April 21, 1917; *Daily Oklahoman*, April 27, 1917.

[11]*Daily Oklahoman*, May 9, 1917.

[12]Blackburn, *Heart of the Promised Land: An Illustrated History of Oklahoma County*, 134.

[13]Board Minutes, March 18, 1918; *Oklahoma News*, September 19, 1918.

[14]Board Minutes, May 3, 1918, March 12, 1919; *Oklahoma News*, September 19, 1918.

[15]Board Minutes, December 27, 1918; *Oklahoma City News*, September 19, 1918.

[16]*Ibid.*

[17]Board Minutes, February 28, 1923; *Oklahoma News*, July 24, 1925.

[18]Board Minutes, November 18, 1924.

[19]Board Minutes, March 20, 1929; *Oklahoma News*, September 29, 1935.

[20]*Daily Oklahoman*, September 25, 1932.

[21]*Daily Oklahoman*, September 21, 1936.

[22]*Oklahoma Farmer Stockman*, October 15, 1925.

[23]*Daily Oklahoman*, September 26, 1919.

[24]*Ibid.*

[25]*Ibid.*

[26]Scrapbook, 1909.

[27]Scrapbook, 1911.

[28]*Oklahoma City Times*, September 27, 1916; *Daily Oklahoman*, September 27, 1916; *Oklahoma News*, September 27, 1916.

[29]*Daily Oklahoman*, May 16, 1920.

[30]Scrapbook, 1908.

[31]*Ibid.*

[32]*Ibid.*

[33]Scrapbook, 1911.

[34]*Oklahoma City Times*, September 28, 1912.

[35]*Oklahoma City Pointer*, September 27, 1917; *Oklahoma News*, September 30, 1917.

[36]*Daily Oklahoman*, September 26, 1921.

[37]*Daily Oklahoman*, September 22, 1929.

[38]*The Employer* (Oklahoma City), (August, 1917), Vol. 2, no. 3, 9.

[39]*Oklahoma City Times*, April 11, 1918.

[40]*Ibid.*

[41]Press release, Scrapbook, 1917.

[42]*Daily Oklahoman*, September 18, 1921.

[43]*Kingfisher Times*, August 9, 1922; *El Reno American*, August 16, 1923; *Edmond Sun*, August 21, 1924; *Shawnee Herald*, September 10, 1925; *Altus Times Democrat*, August 17, 1926; *Stillwater Gazette*, September 16, 1927.

[44]Scrapbook, 1912; *Oklahoma News*, September 28,

1926.

[45]*Daily Oklahoman*, September 24, 1925.

[46]Board Minutes, January 15, 1926; *Oklahoma News*, September 28, 1926.

[47]*Kingfisher Free Press*, August 9, 1929; *Oklahoma News*, September 20, 1932.

[48]Board Minutes, April 8, 1922, February 3, 1923, April 13, 1923.

[49]Harlow, *Makers of Oklahoma Government.*

[50]*Ibid.*

[51]*Oklahoma City Times*, September 22, 1923; Board Minutes, September 18, 1923.

[52]Board Minutes, February 26, 1926.

[53]Board Minutes, November 11, 1924.

[54]Board Minutes, September 29, 1926, October 19, 1926, November 12, 1926, December 17, 1926, January 7, 1927, March 25, 1927, September 9, 1927.

[55]Board Minutes, April 20, 1928, February 6, 1931.

[56]Board Minutes, November 7, 1932, June 14, 1933.

[57]Board Minutes, June 14, 1933.

[58]*Oklahoma City Builder*, September 30, 1936.

[59]*Oklahoma City Times*, undated, Scrapbook, 1936.

[60]*Daily Oklahoman*, July 25, 1937.

[61]Scrapbook 1937.

[62]Scrapbook, 1937.

Chapter Five:

[1]*Daily Oklahoman*, September 26, 1942, September 30, 1942; *Oklahoma City Times*, September 26, 1942.

[2]*Daily Oklahoman*, September 10, 1944.

[3]*Livestock News*, September 5, 1944.

[4]Minutes of the Annual Meeting, November 16, 1944.

[5]Board Minutes, June 11, 1945.

[6]*Daily Oklahoman*, September 16, 1945.

[7]*Ibid.*; Board Minutes, November 3, 1945.

[8]Faulk, Faulk, and Blackburn, *Oklahoma City: A Centennial Portrait.*

[9]*Ibid.*

[10]*Ibid.*

[11]James Smallwood, *Urban Builder: The Life of Stanley Draper* (Oklahoma City: Oklahoma Heritage Association, 1976).

[12]Minutes of the Annual Meeting, November 12, 1946.

[13]*Ibid.*

[14]Board Minutes, September 13, 1948.

[15]*Ibid.*

[16]Board Minutes, October 26, 1948.

[17]Minutes of the Annual Meeting, March 9, 1949; Executive Committee Minutes, March 22, 1950.

[18]Executive Committee Minutes, March 22, 1950; Board Minutes, April 25, 1950.

[19]*Daily Oklahoman*, May 10, 1950.

[20]Minutes of the Annual Meeting, May 22, 1950.

[21]Executive Committee Minutes, March 22, 1950.

[22]Annual Report, 1950.

[23]Minutes of the Annual Meeting, November 14, 1950.

[24]*Ibid.*

[25]Board Minutes, November 13, 1951.

[26]Board Minutes, November 13, 1951.

[27]*Daily Oklahoman*, September 2, 1953.

[28]"Panorama of a Dream: Oklahoma State Fair and Exposition," a promotional booklet prepared by the Pate Organization, in the State Fair of Oklahoma Archives.

[29]*Ibid.*

[30]*Ibid.*

[31]*Ibid.*

[32]*Ibid.*

[33]*Oklahoma City Times*, July 12, 1954.

[34]*Oklahoma City Times*, August 1, 1954.

[35]*Oklahoma City Times*, August 24, 1954.

[36]Interview with Frances Young, February 28, 1989; Annual Report, 1954.

[37]Board Minutes, November 9, 1954.

[38]Annual Report, 1955, 1956.

[39]Executive Committee Minutes, May 10, 1955.

[40]Minutes of a Joint Meeting, January 24, 1956.

[41]Board Minutes, January 16, 1957.

Chapter Six:

[1]Minutes of the Annual Meeting, November 10, 1959; Executive Committee Minutes, January 12, 1960; Minnie Drowatzky, ed., *Who's Who in Greater Oklahoma City* (Oklahoma City: Junior Chamber of Commerce, 1965).

[2]Board Minutes, January 12, 1960; "1960 State Fair Committees," a document in the Minute Books of the State Fair of Oklahoma.

[3]Interview with Edward L. Gaylord, Oklahoma City, June 23, 1993; Letter from C.G. Baker to Executive Committee Members, January 17, 1961; Board Minutes, January 24, 1961.

[4]Interview with E.L. Gaylord, June 23, 1993; Board Minutes, March 20, 1961.

[5]*Sunday Oklahoman*, September 17, 1978.

[6]Board Minutes, March 20, 1961; Interview with Orval "Sandy" Saunders, May 20, 1993.

[7]Annual Report of the State Fair of Oklahoma, 1960 (Annual Reports, published each year by the fair board, will hereafter be referred to as Annual Reports, with the year designated. No page numbers appear in the reports).

[8]Executive Committee Minutes, July 10, 1961; Board Minutes, July 25, 1961; Interview with Sandy Saunders, May 20, 1993.

[9]*Ibid.*; Annual Report, 1961, 1966.

[10]Interview with Sandy Saunders, May 20, 1993.

[11]*Oklahoma City Times*, April 11, 1961; Interview with Sandy Saunders, May 20, 1993; "Panorama of a Dream," an undated brochure published circa 1954 on the plans for a new fairgrounds, located in the State Fair of Oklahoma Archives.

[12]*Daily Oklahoman*, December 6, 1958, October 11, 1961.

[13]*Oklahoma Orbit*, September 17, 1961; *Daily Oklahoman*, September 23, 1961.

[14]*Daily Oklahoman*, September 23, 1961.

[15]Interview with Sandy Saunders, May 20, 1993.

[16]*Oklahoma City Times*, September 2, 1961.

[17]*Daily Oklahoman*, September 23, 1962; Annual Report, 1963; *Oklahoma City Times*, September 28, 1970; September 23, 1971.

[18]*Oklahoma City Times*, October 2, 1964.

[19]Scrapbook, 1959; *Daily Oklahoman*, September 29, 1960; Board Minutes, September 1, 1960.

[20]*Daily Oklahoman*, September 17, 1962; *Oklahoma City Times*, September 23, 1966, September 13, 1967; Scrapbook, 1967; *Daily Oklahoman*, September 25, 1968.

[21]Interview with Sandy Saunders, May 20, 1993; State Fair Program, 1971.

[22]Interview with Sandy Saunders, May 20, 1993.

[23]Executive Committee Minutes, August 31, 1967.

[24]*Oklahoma City Times*, January 18, 1962; Executive Committee Minutes, January 18, 1962.

[25]Board Minutes, October 15, 1962; Executive Committee Minutes, January 18, 1962.

[26]*Oklahoma City Times*, May 17, 1962.

[27]Annual Report, 1963, 1964; Executive Committee Minutes, December 16, 1963; *Daily Oklahoman*, September 26, 1964.

[28]*Daily Oklahoman*, May 17, 1962; *Oklahoma City Times*, September 23, 1964.

[29]Letter from Sandy Saunders to Dean McGee, January 7, 1965; Board Minutes, January 28, 1965.

[30]Board Minutes, November 12, 1968; Executive Committee Minutes, December 12, 1968.

[31]Annual Report, 1964.

[32]Board Minutes, April 29, 1966; *Daily Oklahoman*, September 24, 1966; Executive Commitee Minutes, March 15, 1967; Board Minutes, February 4, 1972; Executive Committee Minutes, July 7, 1972; *Daily Oklahoman*, September 12, 1976.

[33]Annual Report, 1971.

[34]Annual Report, 1976.

[35]Annual Report, 1969, 1970, 1971; *Oklahoma Journal*, September 21, 1967; Executive Committee Minutes, April 10, 1968, January 14, 1971.

Chapter Seven:

[1]Interview with Sandy Saunders, May 20, 1993.

[2]State Fair Program, September 25, 1979.

[3]Interview with Sandy Saunders, May 20, 1993.

[4]*Ibid.*

[5]*Ibid.*; *Oklahoma Journal*, September 18, 1969.

[6]*Daily Oklahoman*, September 27, 1966.

[7]*Oklahoma City Times*, September 14, 1967; *Oklahoma Journal*, September 14, 1968; *Sunday Oklahoman*, September 22, 1968.

[8]*Daily Oklahoman*, September 20, 1969.

[9]*Oklahoma Journal*, September 20, 1975.

[10]Interview with Sandy Saunders, May 20, 1993; Interview with Louise O'Neal, October 1, 1993.

[11]Interview with Sandy Saunders, May 20,, 1993.

[12]Interview with Louise O'Neal, October 1, 1993.

[13]Executive Committee Minutes, March 18, 1965.

[14]*Oklahoma City Times*, September 22, 1965.

[15]State Fair Program, September 26, 1965; *Oklahoma City Times*, September 30, 1965, September 24, 1965, July 26, 1966.

[16]*Daily Oklahoman*, September 29, 1965.

[17]*Ibid.*

[18]Executive Committee Minutes, December 16, 1963, January 29, 1964, February 18, 1964, March 30, 1964; Interview with Sandy Saunders, May 20, 1993.

[19]*Daily Oklahoman*, May 13, 1964.

[20]Annual Report, 1965; Executive Committee Minutes, March 30, 1964; Interview with Sandy Saunders, May 20, 1993.

[21]Annual Report, 1965; *Oklahoma City Times*, September 23, 1964; Interview with Sandy Saunders, May 20, 1993.

[22]Board Minutes, November 13, 1962; *Daily Oklahoman*, April 19, 1967; Executive Committee Minutes, August 31, 1967.

[23]*Daily Oklahoman*, September 28, 1967; Annual Report, 1968.

[24]*Daily Oklahoman*, August 30, 1976; State Fair Program, September 24, 1978.

[25]State Fair Program, September 23, 1979; "The Goodholm Mansion," an undated brochure published by the State Fair of Oklahoma, in the Archives.

[26]Annual Report, 1963, 1965.

[27]Executive Committee Minutes, June 6, 1969; Annual Report, 1969, 1972.

[28]Board Minutes, April 29, 1966; Annual Report, 1966.

[29]Interview with Sandy Saunders, May 20, 1993.

[30]*Ibid.*

[31]Annual Report, 1967; *Oklahoma City Times*, September 24, 1969; *Daily Oklahoman*, September 25, 1971.

[32]Interview with Sandy Saunders, May 20, 1993.

[33]*Daily Oklahoman*, September 28, 1972.

[34]Annual Report, 1971; *Daily Oklahoman*, September 25, 1975.

[35]Annual Report, 1984, 1985.

[36]Annual Report, 1965.

[37]State Fair Program, 1972; Interview with Earl Nichols, Oklahoma City, July 20, 1993.

[38]Interview with Earl Nichols, July 20, 1993; *Daily Oklahoman*, September 27, 1973; State Fair Program, September 24, 1979.

[39]"Sundowners Action," Vol. 3, no. 3 (November, 1993).

[40]Annual Report, 1980.

Chapter Eight:

[1]*Daily Oklahoman*, October 11, 1961, November 22, 1966.

[2]*Oklahoma City Times*, May 17, 1967; *Oklahoma Journal*, February 12, 1968.

[3]Board Minutes, February 12, 1978.

[4]Board Minutes, February 27, 1978.

[5]Memorandum from Joe Rucks to John Parsons, February 24, 1978, in the State Fair Archives.

[6]Minutes of the Annual Meeting, November 15, 1978.

[7]Minutes of the Annual Meeting, November 16, 1979.

[8]Board Minutes, May 13, 1980; *Daily Oklahoman*, May 31, 1980.

[9]Contract between State Fair of Oklahoma and the City of Oklahoma City, July 1, 1980.

[10]Minutes of the Annual Meeting, November 24, 1981; KWTV newscast, July 7, 1981.

[11]Report from Sandy Saunders to Board of Directors, June 17, 1981.

[12]KWTV newscast, July 7, 1981.

[13]Contract between State Fair of Oklahoma and the City of Oklahoma City, July 1, 1980, Article IV, section 4.1b.

[14]Board Minutes, February 10, 1982, May 7, 1982; *Daily Oklahoman*, May 12, 1982.

[15]Board Minutes, August 12, 1982; Contract between the State Fair of Oklahoma and the City of Oklahoma City, August 3, 1982.

[16]Letter from Councilman Franks to Mayor Andy Coats, December 7, 1983, in the archives of the State Fair of Oklahoma.

[17]Lease Agreement between the City of Oklahoma City and the Fairgrounds Trust, October 3, 1983; Oklahoma City Fairgrounds Trust Indenture, October 3, 1983.

[18]Board Minutes, August 12, 1982, November 18, 1982; Proposal to the State Fair Board of Directors, November 18, 1983.

[19]Board Minutes, February 10, 1983.

[20]Board Minutes, May 12, 1983, August 9, 1983, September 28, 1983.

[21]Board Minutes, August 14, 1984.

[22]Executive Committee Minutes, April 5, 1985.

[23]Annual Report, 1985.

[24]Board Minutes, November 19, 1985.

[25]Interview with Donald J. Hotz, September 18, 1993.

[26]*Ibid.*

[27]Odie B. Faulk, *Oklahoma: Land of the Fair God* (Los Angeles: Windsor Publications, 1986), 253.

[28]Board Minutes, August 13, 1986.

[29]Interview with Donald Hotz, September 18, 1993.

[30]Annual Report, 1987.

[31]*Daily Oklahoman*, September 27, 1988.

[32]Interview with Donald Hotz, September 18, 1993.

[33]Annual Report, 1987, 1993.

[34]*Daily Oklahoman*, December 2, 1958, February 15, 1959, April 2, 1960.

[35]*Daily Oklahoman*, July 18, 1961.

[36]*Oklahoma City Times*, May 3, 1965; *Daily Oklahoman*, November 25, 1965.

[37]*Daily Oklahoman*, September 12, 1965.

[38]*Daily Oklahoman*, May 25, 1966; *Oklahoma City Times*, May 25, 26, 1966.

[39]"List of Horse Shows on the Fairgrounds, 1993," in the files of the State Fair of Oklahoma.

[40]Interview with Donald Hotz, September 18, 1993.

[41]Interview with Donald Hotz, September 18, 1993.

[42]*Ibid.*

[43]Board Minutes, November 16, 1988.

[44]*Sunday Oklahoman*, September 17, 1989; Annual Report, 1989.

[45]Annual Report, 1993.

[46]Board Minutes, August 19, 1992.

[47]State Fair Press Release, October 3, 1993.

[48]*Ibid.*

[49]Letter from Don Hotz and Edward C. Joullian, III, to Ronald J. Norick, July 2, 1993.

[50]*Ibid.*

[51]Annual Report, 1993.

INDEX

A&M College, 65, 68, 120
A&M College Building, 60
Administration Building, 57
Agricultural Displays, 164, 218
Agricultural School, 66, 84
Agricultural Society of the Cherokee Nation, 5
Agriculture Building, 58, 60, (photo of) 125, 130, 154, 181
Agriculture and Youth Expo, 218
Air Conditioning, 157
Air Corps Cadet Training Program, 112
Air Dome Theater, 60
All-Arabian Horse Show, 212
American Horse Show Association, 174
Amusement Corporation of America, 166
Anderson, Vincent "Farmer", 44
Anthony, C.R., 139
Appliances Building, (photo of) 126, 130, 181
Arena, 154, 155, 156, (photo of) 156, 157, 172, 173, 210, 211, 220
Arrows to Atoms Tower, 155, 178
Art Center and Planetarium, (photo of) 145
Art Department, 44
Arts and Crafts Building, 157
Auto Polo, 71
Auto Show, 91, 92
Automobile and Carriage Building, 60, 91
Automobile Racing, 70, 72, 73, 105, (photo of) 172
Aviation, 92, 96

B., Ida, (photo of) 171
Baker, C.G. "Pete," 121, 123, 133, 137, 138, 143
Bassett, Mary Jane, (photo of) 211
Bath, Vincent L, 36
Bee and Honey Building, 60
Beef Cattle Competition, 164
Belle Isle Amusement Park, (photo of) 24
Bennett, Clay, 217
Bennett, Henry G., 120
Beutler, Jiggs, (photo of) 143
Beutler and Son, 174
Bicentennial Celebration, 159, 169
Bishop, Bill, 197
Bixby, Tams, 76
Blankenship, G.T., 201
Blazers, 211
Bond Issue, 125
Boy Scout Headquarters, 134
Browne, Virgil, 139

Cain, William Morgan, 129
Canals of Venice, 59
Carnival Games, 170
Carousel, 59, 77
Cement Exhibit Building, 60
Centennial Commission, 216
Chamber of Commerce, Oklahoma City, 32, (photo of) 32, 33, 35, 61, 103, 104, 106, 108, 116-118, 124, 128, 180, 182, 201, 217, (see also Commercial Club)
Champion, Frank, 94
Cherokees, 5
Children's Barnyard, 154
Citizens Advisory Committee, 108
City of Oklahoma City, 119
Civil Aeronautics Administration, 150
Classen, Anton, (photo of) 18
Clock Tower, 161

Coady, Edward, 41
Coats, Andy, 199
Colcord, Charles, 26, (photo of) 35, 36, 47
Commercial Club, 12, 15, 32, (see also Chamber of Commerce)
Commercial Club Fair and Exposition, 13
Concession Hall, 77
Constitutional Fountain, (photo of) 158, 160
Cook, Jim, 191
Cooking Department, (photo of) 190
Cornett, Jack, 197, 198
Coston & Frankfort, 126
County Agricultural Extension Agent, 66, 67
County Exhibits, 44
Coyle, William H., 23
Crafts, (photo of) 170
Cruce, Governor Lee, 70, 75, 76

Dairy Barn, 155
Dairy Exhibit, 164
Dallas Plan, 80, 82
Dallas State Fair, 37, 41
Davis, George A., (photo of) 128
Dawkins, Paul, 208
Dee, Tom J., 120, (photo of) 127
Deitz, Oscar, 126
Delmar Gardens, (photo of) 23, (photo of) 25, 36
Dickerson, Leonard, 129
Disbrow, Louis, (photo of) 73
Disney On Ice, 210, (photo of) 212
Douglas High School, 123
Drag Strip, 211
Draper, Stanley, 104, 118, 120, 121, 128, 140
Dulaney, Luther, 139

Earp, Ancel, 120, 124
Economic Development Foundation, 201
89ers Baseball Team, 86, 211
Emerson, Lt. Arthur, 96
Entrance Building, (photo of) 85
Everest, J.H., (photo of) 80
Exhibit Building, 57
Exposition Building, 42, 81

Farm Implement Show, 88
FFA Building, 113, 130, (photo of) 133
FFA Junior Livestock Show, 217
50 States Flag Plaza, 158
Figure "8," 59
First National Bank & Trust Co., 133
Food Service, 170
Ford, President Gerald, 169
Fort Sill Bomber Fleet, 95
4-H Boys and Girls, 113, 130, 217
4-H Building, 86, 87, 113
14 Flags Plaza, 155
Frank, Robert M., 196
Free Fairs Bill, 66
Frizzell Coach Works, (photo of) 157
Future Farmers of America, 87, 113 (see also FFA Building)

Games, 169
Garden and Flowers Building, 178, (photo of) 187
Garden and Flower Show, 178
Garrett, Sandy, (photo of) 211
Gaylord, E.K., 76, 102

Gaylord, Edward L., (photo of) 140, 141-143, (photo of) 164, 190, 194, 202
Geary, James, 13, 15, 16
General Exhibits Building, (photo of) 124, 129
Gerrer, Father, 45
Gilbert, Jerry, 194
Gill, Bill Jr., 134, (photo of statue) 202
Goodale Dirigible, 94
Goodholm Mansion, 179
Grandstand, 57, 59, 77, 81, 84, 129, 172
Greater Oklahoma City Committee, 124
Greater Oklahoma City, Inc., 117
Greer, Frank, 23
Guthrie, 23

Hall, Boots, 217
Haskell, Governor Charles N., 3, 42
Hemphill, Ralph T., 97, 101, 102
Hill, L.L., 93
Historical Spectaculars, 99
Home Arts and Crafts Building, 155
Horse Barns, 155, 161, 212-213
Horse Exercise Arena, 161
Horse Industry, 154
Horse Racing, 18, 47, 69, 70
Horse Shows, 69, 105, 213-214
Hospitality Club, 170
Hot Air Balloon, 93
Hotz, Donald J., 204-206, (photo of) 207
Hubert Castle Circus, 173
Hulsey, William, (photo of) 194, 196

Ice Capades, 173, (photo of) 174, 210
Ice Skating Show, 172
Implement Dealers Day, 67
Implement Row, 67
Independence Arch, (photo of) 158, 160
Indian Building, 85
Indian Exhibits, (photo of) 51
Interim Division, 214
Interim Operations, 105, 133-135, 198, 211, 214
International Building, 181, 186
International Exhibit, 181-186
International Visitors Council of the Chamber, 186

James, Dan, 140
James, Goree, 197, 201
Jeffords, T.M., 65
Jones, C.G., 12, (photo of) 12, 13, 15, 22, 35, 36, 42, 54
Jones, Lem, 124, 145
Jones, Reba, 215
Jordan, Dr. Francis M., 36
Joullian, Edward C., III, 206, (photo of) 206, 207, (photo of) 211, 214, 216, 221
Junior Hospitality Club, (photo of) 197

Keel, John, (photo of) 189
Keller, A.H., 36
Kennedy, Donald, 139
Kilpatrick, John Jr., 139
Kimura, Kimio, 178
Kirkpatrick, John & Eleanor, 139, (photo of) 145, 147
Kirkpatrick, Art Center, 147
Kirkpatrick Planetarium, 151
Kirkpatrick Science and Arts Museum, 147

Lamb, Floyd S., (photo of) 102
Landscaping, 157
Last Frontier Council of the Boy Scouts, 211
Layton, Solomon, 83
Lee, Oscar G., 91
Liberal Arts Building, 81, (photo of) 82, 85

Link Carnival, (photo of) 194
Link, Rod, 166
Little, Phil, 170
Livestock Department, 44, 60
Livestock Pavilion, 60, 81, 113, 120, 130, (photo of) 164
Livestock Shows, 87, 164
Louis Disbrow Racing Team, 70
Luminary Project, 159

Macklanburg, Louis A., 120, 122, 124, 128, (photo of) 129, 129, 139
Made in Oklahoma Building, 160, 178, 180, 181
Mahan, I.S. "Dick," 54, (photo of) 57, 101
Massad, Major Gen. E.L. "Mike", (photo of) 204, 206
McClure, William, 16, 17
McGee, Dean A., 139, (photo of) 150, (photo of) 172, 178, 194
McKeand, Alexander W., 35
McSpadden, Clem, 175
Merchants' Trades Carnival, 10
Metropolitan Area Projects, 219-221
Midway, 46, (photo of) 101, (photo of) 103, 165, (photo of) 217
Military Exhibits, 148-153
Milking Parlor, 161
Mineral Resources Building, 60, 100
Monorail, 177-178, (photo of) 177, (photo of) 193
Montgomery, John, 202
Motor Style Show, 92
Motor Truck Industry, 88, 90
Motorcycle Racers, (photo of) 81
Music Hall, 42, 57
Muskogee, (photo of) 4, 6, 73
Muskogee Fair Bill, 76
Muskogee Indian International Fair, 6
Muskogee State Fair, 73-76, 79

National Air and Space Administration, 151
National Finals Rodeo, 212
Nelson Morris and Company, 61
New Orleans Fair, 203
Nigh, Lt. Gov. George, (photo of) 181
Noble, Rusty, 217
Nofsgar & Lawrence, 126
Norick, Mayor Ronald, 219

O'Neal, Fred & Louise, 171
O'Neil, John E., (photo of) 86
Oklahoma A&M College, 44, 64, 87, 119, 126, 128
Oklahoma A&M College Building, 65
Oklahoma City Arts Center, 135
Oklahoma City Branch of Oklahoma State University, 120
Oklahoma City Charity Horse Show, 212
Oklahoma City Fair and Race Meeting, 26
Oklahoma City Home Show and Furniture Fair, 211
Oklahoma City Is First Fair, 8-9
Oklahoma City Oil Field, 107
Oklahoma City Parks Department, (photo of) 159
Oklahoma City Public Property Authority, 195
Oklahoma City Public Schools, 119
Oklahoma City Round-Up Club, 134
Oklahoma City Science and Arts Foundation, 135
Oklahoma County Free Fair, 114
Oklahoma Fair Park Amusement Company, 59
Oklahoma Geological Survey (photo of) 56
Oklahoma Industries Authority, 117
Oklahoma Industries, Inc., 116
Oklahoma Publishing Company, 141
Oklahoma Publishing Company Building, (photo of) 127, (photo of) 128, 130
Oklahoma Society of the Daughters of the American Revolution, 179

Oklahoma Territorial Fair Association, 15
OSU Technical Institute, 120
Outdoor Lighting, 57
Overholser, Ed, 13, 16, (photo of) 17, 22, 104
Overholser, Henry, (photo of) 8, (photo of) 37, 39, 52, 53,
    54, 56, 76, 77, 81
Owen, J.M., (photo of) 92, 126
Owen, Nelson, 140

Parimutuel Machines, 70
Parking Lots, 106
Parr & Adderhold, 126
Parsons, John, 191
Permanent Stalls, 161
Pirtle, Queen, 54
Pirtle, Bill, 217
Police & Firefighters Academy, 128
Poultry Building, 57
Poultry Fair, 10
Powell, Walter, 193, 194
Preparation Day, 112
Promotions Division, 215
Pronto Pup, 172
Pullman Porters' Quartet, (photo of) 85
Putnam, I.M., 36

Radios, 100
Radio Building, 100
Radio Exhibit, 100
Radio Show, 100
Rainey, Judge Robert M., (photo of) 125
Rea, Frank, 93
Richardson, D.C., 16
Riley, F.M., 15
Ringling Brothers Barnum & Bailey Circus, 210, (photo of)
    218
Robinson's Racing Pigs, (photo of) 198
Rodeo, 174
Rodeo Cowboys' Association, 174
Royal American Shows, 165, 166
Rucks, Joe, 191, 194

Sales Facility, 221
Sand Sculpture Project, 208
Saunders, Orval O., "Sandy," (photo of) 140, 143-144, 194,
    202, 204
Schwartz, Walter, 208
Schwartzchild and Sulzberger, 61
Science and Arts Museum, (photo of) 145
Scientific Farming, 63, 68
Scott, Jack L., and Associates, 155
Scott, Jim, 197
Scout-O-Rama, 211
Seattle World's Fair, 178
Second Annual Street Fair, 26
Semicentennial Exposition, 133
Shelley, Frank H., 26, 36
Ship by Truck, 89
Shooting Gallery, (photo of) 90
Sidewalks and Street Paving, 57, 155, 158
Sirloin Club of Oklahoma, 217
Sites, J.T., 54
Sooner's Trail, 46

Sorey, Hill & Sorey, 126
Southwest American Livestock Foundation, 217
Southwest Boat Show, 211
Space Tower, 178
State Fair Agricultural School, 64, 65
State Fair Association of Oklahoma, 36, 47
State Fair of Texas, 38, 80
State Fair Committee of the Oklahoma City Chamber of
    Commerce, 185
State Fair Division, 214
State Fair Rodeo, 174, 209
State Fair Stockade, 160
Stiles, Captain Daniel, 13, 16
Stock Car Racing, 211
Storm, W.W., 24
Strasbaugh, Paul B., 194, 201, 202
Street Fair, 24
Street, Mayor Allan, 134
Strobel, Charles J., 94
Sullivan, Marie, 145
Sundowners, 185, 186
Swisher, Bill, 194, 198, (photo of) 200

Tahlequah, 5
Teacher of the Year Award, 216
Teen Fair, 167, 169
Territorial Fair of 1892, (photo of) 13
Territorial Fair Association, 23
Ticket Master, 215
Ticketing Division, 215
Tinker Air Force Base, 112, (photo of) 141, 151-153
Tinker Air Force Base Memorial Aircraft Exhibit, 215
Togni, Flavio, (photo of) 209
Transportation Building, 154, 160
Travel and Transportation Building, 186
Treat Engineering Co., 124
Treat, Guy, 126
Trosper, H.G., 16

United States Department of Commerce, 182
United States Government War Committee on Conventions,
    113
Universal Designs, 177
USDA, 68

Valberg, Walter, 126
Vose, Charles, 194

Walton, Gov. John "Jack", 102
Warren, J.F., (photo of) 76
Wichita State Fair, 41
Wilbur, Phil, 125
Williams, Gov. Robert L., 66
Wolf, Larry, 140
Women's Building, 60, (photo of) 81, 130, 154, 155
Women's Committee of the Oklahoma City Symphony, 179
Women's Rest Cottage, 81, 83
World War I, 97-99, 111-115
World War II, Impact on the Fair, 111-115
World's Fair, 201
Wright & Selby, 126

Young, Frances, 145, 194